Frontiers of Land and Water Governance in Urban Regions

A society that intensifies and expands the use of land and water in urban areas needs to search for solutions to manage the frontiers between these two essential elements for urban living. Sustainable governance of land and water is one of the major challenges of our times. Managing retention areas for floods and droughts, designing resilient urban waterfronts, implementing floating homes, or managing wastewater in shrinking cities are just a few examples where spatial planning steps into the governance arena of water management and vice versa. However, water management and spatial planning pursue different modes of governance, and therefore the frontiers between the two disciplines require developing approaches for setting up governance schemes for sustainable cities of the future. What are the particularities of the governance of land and water? What is the role of regional and local spatial planning? What institutional barriers may arise? This book focuses on questions such as these, and covers groundwater governance, water supply and wastewater treatment, urban riverscapes, urban flooding, flood risk management, and concepts of resilience. The project resulted from a summer school by the German Academy for Spatial Research and Planning (ARL) organized by the editors at Utrecht University in 2013.

This book was originally published as a special issue of *Water International*.

Thomas Hartmann is Assistant Professor in the Department of Human Geography and Spatial Planning at Utrecht University, The Netherlands. His research looks at the relationship between land and water, with a focus on river floods and retention. In this research, he combines planning theory, law and property rights, and water governance.

Tejo Spit is Professor in the Department of Human Geography and Spatial Planning at Utrecht University, The Netherlands. He specialises in land policy, planning methodology, infrastructure planning, and administrative aspects of spatial planning. He has worked both in the academic world and the more problem-oriented world of municipalities.

Routledge Special Issues on Water Policy and Governance

Edited by:

Cecilia Tortajada (IJWRD) – *Institute of Water Policy, National University of Singapore*

James Nickum (WI) – *International Water Resources Association, France*

Most of the world's water problems, and their solutions, are directly related to policies and governance, both specific to water and in general. Two of the world's leading journals in this area, the *International Journal of Water Resources Development* and *Water International* (the official journal of the International Water Resources Association), contribute to this special issues series, aimed at disseminating new knowledge on the policy and governance of water resources to a very broad and diverse readership all over the world. The series should be of direct interest to all policy makers, professionals and lay readers concerned with obtaining the latest perspectives on addressing the world's many water issues.

Water Pricing and Public-Private Partnership
Edited by Asit K. Biswas and Cecilia Tortajada

Water and Disasters
Edited by Chennat Gopalakrishnan and Norio Okada

Water as a Human Right for the Middle East and North Africa
Edited by Asit K. Biswas, Eglal Rached and Cecilia Tortajada

Integrated Water Resources Management in Latin America
Edited by Asit K. Biswas, Benedito P. F. Braga, Cecilia Tortajada and Marco Palermo

Water Resources Management in the People's Republic of China
Edited by Xuetao Sun, Robert Speed and Dajun Shen

Improving Water Policy and Governance
Edited by Cecilia Tortajada and Asit K. Biswas

Water Quality Management
Present Situations, Challenges and Future Perspectives
Edited by Asit K. Biswas, Cecilia Tortajada and Rafael Izquerdo

Water, Food and Poverty in River Basins
Defining the Limits
Edited by Myles J. Fisher and Simon E. Cook

Asian Perspectives on Water Policy
Edited by Cecilia Tortajada and Asit K. Biswas

Managing Transboundary Waters of Latin America
Edited by Asit K. Biswas

Water and Security in Central Asia
Solving a Rubik's Cube
Edited by Virpi Stucki, Kai Wegerich, Muhammad Mizanur Rahaman and Olli Varis

Water Policy and Management in Spain
Edited by Francisco González-Gómez, Miguel A. García-Rubio and Jorge Guardiola

Frontiers of Land and Water Governance in Urban Regions

Edited by

Thomas Hartmann and Tejo Spit

Routledge
Taylor & Francis Group

LONDON AND NEW YORK

IWRA

First published 2016
by Routledge
2 Park Square, Milton Park, Abingdon, Oxon, OX14 4RN, UK

and by Routledge
711 Third Avenue, New York, NY 10017, USA

Routledge is an imprint of the Taylor & Francis Group, an informa business

British Library Cataloguing in Publication Data
A catalogue record for this book is available from the British Library

ISBN 13: 978-1-138-91115-4

Typeset in Times
by RefineCatch Limited, Bungay, Suffolk

Publisher's Note
The publisher accepts responsibility for any inconsistencies that may have arisen during the conversion of this book from journal articles to book chapters, namely the possible inclusion of journal terminology.

Disclaimer
Every effort has been made to contact copyright holders for their permission to reprint material in this book. The publishers would be grateful to hear from any copyright holder who is not here acknowledged and will undertake to rectify any errors or omissions in future editions of this book.

Contents

TABLE OF CONTENTS

Citation Information

The chapters in this book were originally published in *Water International*, volume 39, issue 6 (October 2014). When citing this material, please use the original page numbering for each article, as follows:

Chapter 1
Editorial: Frontiers of land and water governance in urban regions
Thomas Hartmann and Tejo Spit
Water International, volume 39, issue 6 (October 2014) pp. 791–797

Chapter 2
Groundwater governance and spatial planning challenges: examining sustainability and participation on the ground
Gabriela Cuadrado-Quesada
Water International, volume 39, issue 6 (October 2014) pp. 798–812

Chapter 3
Impact of short-rotation coppice on water and land resources
Jens Hartwich, Jens Bölscher and Achim Schulte
Water International, volume 39, issue 6 (October 2014) pp. 813–825

Chapter 4
Regional governance vis-a-vis water supply and wastewater disposal: research and applied science in two disconnected fields
Martin Schmidt
Water International, volume 39, issue 6 (October 2014) pp. 826–841

Chapter 5
Managing urban riverscapes: towards a cultural perspective of land and water governance
Meike Levin-Keitel
Water International, volume 39, issue 6 (October 2014) pp. 842–857

Chapter 6
The governance dilemma in the Flanders coastal region between integrated water managers and spatial planners
Karel Van den Berghe and Renaat De Sutter
Water International, volume 39, issue 6 (October 2014) pp. 858–871

Please direct any queries you may have about the citations to
clsuk.permissions@cengage.com

Notes on Contributors

Jens Bölscher is a Researcher at the Institute of Geographical Science, Applied Geography – Environmental Hydrology and Resource Management, Freie Universität Berlin, Germany.

Antje Bruns is a Junior Professor of Climate Change and Sustainable Development at the Institute of Geographical Science, Freie Universität Berlin, Germany.

Gabriela Cuadrado-Quesada is a PhD candidate in the Faculty of Law at the University of New South Wales, Australia. Her doctoral research focuses on groundwater governance. She has previously completed a Masters in Water and Coastal Management and Environmental and Infrastructure Planning at the University of Oldenburg, Germany, and at the University of Groningen, The Netherlands. Her thesis investigated the implementation of international treaties in integrated coastal management in Costa Rica, Germany, and South Africa.

Renaat De Sutter is a Visiting Professor in the Department of Civil Engineering at the University of Ghent, Belgium.

Thomas Hartmann is Assistant Professor in the Department of Human Geography and Spatial Planning at Utrecht University, The Netherlands. His research looks at the relationship between land and water, with a focus on river floods and retention. In this research, he combines planning theory, law and property rights, and water governance.

Jens Hartwich is a Researcher at the Institute of Geographical Science, Applied Geography – Environmental Hydrology and Resource Management, Freie Universität Berlin, Germany.

Karen Hetz is based in the Department of Geography, Humboldt Universität zu Berlin, Germany.

Meike Levin-Keitel is a Researcher at the Institute of Environmental Planning, Leibniz University of Hanover, Germany.

Guy A. Meadows is Director of the Great Lakes Research Centre, and Adjunct Professor of Geological and Mining Engineering and Sciences at Michigan Technological University, Houghton, MI, USA.

Richard K. Norton is Chair and Associate Professor of the Urban and Regional Planning Program at the University of Michigan, Ann Arbor, MI, USA.

NOTES ON CONTRIBUTORS

Martin Schmidt is a Researcher in the Department of Spatial and Infrastructure Planning at the Technische Universität Darmstadt, Germany.

Achim Schulte is a Professor at the Institute of Geographical Science, Applied Geography – Environmental Hydrology and Resource Management, Freie Universität Berlin, Germany.

Tejo Spit is Professor in the Department of Human Geography and Spatial Planning at Utrecht University, The Netherlands. He specialises in land policy, planning methodology, infrastructure planning, and administrative aspects of spatial planning. He has worked both in the academic world and the more problem-oriented world of municipalities.

Barbara Tempels is a PhD student at the Centre for Mobility and Spatial Planning, Ghent University, Belgium.

Karel Van den Berghe is a PhD student at the Centre for Mobility and Spatial Planning, Ghent University, Belgium.

Frontiers of land and water governance in urban regions

A society that intensifies and expands the use of land and water in urban areas needs to rethink the relation between spatial planning and water management. The traditional strategy to manage land and water under different governance regimes no longer suits the rapidly changing environmental constraints and social construction of the two key elements in urban development. The dynamics of urban development and changing environmental constraints cause an urgent need for innovative concepts in the overlapping field of land and water governance.[1] The claim for more space for rivers for flood retention (Hartmann, 2011) and environmental protection (Moss & Monstadt, 2008), the fragmentation of the drinking water sector (Moss, 2009), or unsolved upstream–downstream relations (Scherer, 1990) are illustrative of these dynamics. Therewith, increasingly, water management steps into the governance arena of spatial planning, and spatial planning needs to reconsider its notions of water issues.

Particularly in urban regions, engineering and technical solutions of water management reach their boundaries; new frontiers for the common governance of land and water emerge (Figure 1). Although agriculture remains important for land and water governance (Calder, 2005), and it is the biggest consumer of water and occupies large areas of land, this special issue focuses on the urban realm because in the tense relation between water and land, the need for innovative approaches is more urgent. Urban regions are intensively used by many different stakeholders with competing interests, so that frictions between socio-economic dynamics and environmental constraints of land and water are more complex and more intense. Hence, the challenges of finding creative and path-breaking solutions in those areas are most pressing.

Resolving and dealing with such tensions and frictions asks for a reconsideration of the traditional institutional divide between spatial planning and water management. New governance schemes need to be found that are complementary to those traditional institutions. Water management and spatial planning usually pursue essentially different modes of governance: water management traditionally relies on engineering and technical solutions, spatial planning usually mediates between competing interests without having its own strong institutional capacities (Hartmann & Driessen, 2013; Moss, 2004). Spatial planning is thereby more comprehensive and meta-disciplinary than water management, which tends to be more specific and sectoral (Moss, 2009). Whereas water engineers aim to control and regulate the water sector, spatial planning aims for the coordination and integration of many different sector activities (Hartmann & Juepner, 2014).

There is an ongoing academic discussion on connecting and integrating sectors and subsectors in the field of water management (Dyckman & Paulsen, 2012; Gleick, 2000; Wiering & Immink, 2006). What exactly integrated water resource management (IWRM) means, and what should be integrated in what, remains vague (Biswas, 2004). But a call to involve other disciplines has been issued, in a much broader context than just water management (Loucks, 2000). Others promote an integration of 'natural systems' (water and land) in

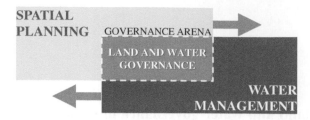

Figure 1.   Emerging and increasing field of land and water governance.

the 'human systems', involving economy, policy, institutions and others (Jønch-Clausen & Fugl, 2001). Calder notes in *Blue Revolution* that the 'revolution in the way land and water are managed' (Calder, 2005, p. 1) is a philosophical one, changing the way society regards water. He acknowledges the need to invent governance schemes to deal with this revolution. However, his investigation of integrated land and water resources management focuses on forests and agricultural aspects. Another debate centres on legal integration of water issues in order to create more comprehensive water laws (Hartmann & Albrecht, 2014; Gilissen, van Rijswick, & van Schoot, op. 2009; Heer et al., 2004; Jong, 2007).

Edelenbos et al. pursue a broader perspective on water governance, including also Delta regions, spatial planning. They discuss the connective capacity of water (Edelenbos, Bressers, & Scholten, 2013b), and state in their conclusion that water governance systems are 'complex, multilevel systems, heavily intertwined with other physical, social, political and economic subsystems' (Edelenbos, Bressers, & Scholten, 2013a, p. 333). They call for continued work on resolving the fragmentation in the water sector, acknowledging that connection is not the ultimate answer to water governance (p. 344). Billé identifies four common illusions and misbeliefs of integrating and connecting the water sector with other environmental governance by referring to integrated coastal zone management (ICZM) (Billé, 2008). He shows, first, that governance problems are not solved by bringing all stakeholders to a round table; second, there is not one manager; and third, the public interest is not easily identifiable, and more knowledge does not necessarily solve governance problems ('positivist illusion'). Governance schemes instead need to respond to the specific governance problems; or as Edelenbos et al. phrase it: water governance needs to be 'aware of the problems with existing boundaries in water governance' (Edelenbos et al., 2013a, p. 349). Often, approaches focus instead on the institutional boundaries of land and water management (Grigg, 2008).

Environmental constraints and new governance frontiers

Frontiers of land and water governance emerge along the physical boundaries between land and water: along *horizontal* boundaries on riverfronts and coastlines, along *vertical* boundaries between groundwater or water infrastructure and surface land use, and along *fluid* boundaries in floodplains and due to changing sea levels. Figure 2 illustrates the

VERTICAL   HORIZONTAL   FLUID

Figure 2.   Frontiers of land and water governance.

three frontiers schematically. Along these physical boundaries, land and water issues overlap in the urban realm and create new governance frontiers.

In order to make the analytical concept of 'governance frontiers' clear, it is vital to distinguish between a frontier and a boundary. A boundary is something that indicates or fixes a limit or extent; a frontier comes with different meanings: it is a region that forms the margin of settled territory, the farthermost limits of knowledge, a division between different or opposed things, or a new field for developmental activity. So, on the one hand, frontiers of land and water governance refer to land and water as opposed to each other; but more importantly, they also express that the common governance of land and water is a field for development and research.

Vertical frontiers

The vertical boundary between land and water is between groundwater and water infrastructure below and land use above.[2] Often the interactions between urban land uses on top and water below, like drinking water supply, pipes for freshwater and sewage that enable land uses, or pollution of ground water, occur quite unnoticeably. The surface is the boundary between water below and land on top. However, socio-economic and environmental changes challenge this boundary in various ways. Three contributions in this special issue show the scope of governance challenges on this vertical frontier.

Cuadrado-Quesada explores, with examples from South Australia and Costa Rica, how land uses and groundwater interfere and what in these cases are the specific governance challenges. This paper therewith combines an analysis of the regulatory framework with the issue of participation.

Hartwich, Bölscher and Schulte present a very specific example of a vertical governance frontier between land and water. Short rotation coppices reduce the groundwater level significantly during the growth period (Perry, Miller, & Brooks, 2001). From a governance perspective this raises questions about how to allocate and distribute the advantages and disadvantages of those effects. Hartwich et al. analyse in particular the dimensions of efficiency and effectiveness.

Schmidt connects regional governance and water supply and wastewater disposal in three different case studies in Germany: Berlin, the Ruhr and Frankfurt (Main). This contribution comprises an introduction into how Germany organizes the two fields of governance and regional planning and water management separately. Schmidt identifies the inherent need to address land and water with a comprehensive governance approach.

The three contributions in this section of the special issue illustrate that the vertical boundary between land and water can become more permeable because of the various and not always obvious interdependencies between land use on top and water issues below. In order to find appropriate governance schemes for this frontier, one needs to tackle problems that are not visible in the first place and situations of externalities and long-term effects. This will most likely make it more difficult to activate stakeholders and reach commitment within a certain governance arrangement.

Horizontal frontiers

The horizontal boundary of land and water establishes itself along riverfronts and coastlines. In terms of governance, such areas are usually contested terrains: tourism, environmental protection, real estate development and others issues are in conflict with regard to the use of land and water. In fact, analytically one can observe various governance frontiers between spatial planning and water management established along waterfronts. Two contributions discuss the governance challenges at the horizontal frontier at rivers and the coastline.

Levin-Keitel offers a cultural perspective on urban riverscapes. Using two examples from southern Germany – the cities of Ratisbon and Nuremberg – she searches for the role of culture (i.e. norms and values) in integrated planning processes of urban riverscapes and discusses the dynamic cultural complexes when dealing with urban riverscapes. Levin-Keitel illustrates different cultural imprints of water governance and land governance. This helps in understanding the challenges (and need) for a more integrated riverscape management.

Van den Berghe and De Sutter present a very interesting and very specific horizontal frontier of land and water governance: the case of Flanders' coastal region. Their analysis helps one to understand how effective and efficient coastal management is challenged by historically and geographically established lock-in situations.

Finally, there are many examples of extending horizontal frontiers of land and water governance in urban regions that show the increasing intensity and importance of developing appropriate governance schemes for those areas. Conflicts of interest result from economic and ecological functions of waterways and shorelines competing with interests of land-use planning. The complexity of such problems has been addressed previously in discussions around ICZM (Billé, 2008) or marine spatial planning, but, particularly in urban areas, governance schemes need to address the increasing socio-economic (e.g., waterfront development projects, floating homes, etc.) and environmental dynamics (e.g., sea level rise, water quality, etc.) along the physical boundary between water and land.

Fluid frontiers

Fluid frontiers between land and water governance refer to situations where the physical boundary between land and water is changing permanently or temporarily (Brown & Damery, 2002; Hartmann & Spit, 2012). This predominantly is the case with storm surges, sea-level rise, but also with the desiccation of lakes (e.g., Aral Lake). In addition, climate change affects water issues in many ways and diminishes or changes boundaries between land and water, which calls for new governance solutions. Fluid boundaries between land and water question existing governance schemes in specific ways. The most prominent fluid boundary in urban regions is certainly flooding, especially because many urban areas are located on large water bodies (Hartmann, 2011). Three contributions address the fluid frontier from different methodological perspectives, covering examples from different continents.

Tempels and Hartmann reflect on this frontier through the lens of the concept of co-evolution. This is a theoretical contribution on the trade-off of flexibility with robustness in flood risk management. The stance taken by Tempels and Hartmann is that floods need to be regarded not as purely technical water management issues, but rather in the context of interacting (i.e. co-evolving) systems: namely land and water.

Hetz and Bruns apply the theoretical concept of lock-in to urban planning in Johannesburg. They discuss informal growth, land policies and how they interact with urban flooding. Thereby issues such as distributional injustice are addressed. The authors provide insights into the constrains with regard to adaptive planning for climate change.

Norton and Meadows present a case where different governance frontiers interact. At the Great Lakes in the United States and Canada, tidal movements affect land uses on the shorelines. Norton and Meadows explain the complex institutional and legal consequences of this interaction of land and water. This case is specific for the Great Lakes, but the effect on affluent land (in a literal sense) is also of interest for regions that will suffer from sea-level rise.

These three contributions explain why water management but also spatial planning cannot rely on established and well-rehearsed procedures and institutions in the face of fluid frontiers. Instead, new governance schemes are required (Dworak & Görlach, 2005). Entrenched lock-in situations need to be overcome (Wesselink, 2007).

Governance challenges frontiers of land and water governance

The above distinction of vertical, horizontal and fluid governance frontiers of land and water is an analytical framework that reveals the different governance challenges. Whereas the horizontal frontier has to deal with governance problems that are very long-term and rather invisible, or, at least where causes and effects are not always obvious, the vertical frontier is one of high levels of socio-economic and environmental dynamics. The governance challenges along the fluid frontier, however, need to overcome entrenched lock-in situations and deal with uncertainty and normativity of flood-risk perceptions in a particular way. The examples illustrate the scope and urgency of these frontiers and the urgency for a quest for innovative governance approaches to address these frontiers.

Acknowledgement

This special issue is an outcome of the international academic summer school of the German Academy for Spatial Research and Planning (Akademie für Raumforschung und Landesplanung, ARL). It took place in August 2013 and assembled a number of PhD students and professors, all active in research on the issue of governance of land and water. Most contributions in this special issue stem from this summer school, which was organized at Utrecht University by the two guest editors. The guest editors and authors are grateful that *Water International* supported them in bringing their efforts together in this special issue, in particular because it provided many younger authors a chance to publish their work.

Thomas Hartmann and Tejo Spit
Urban and Regional Research Centre Utrecht (URU),
Utrecht University, Utrecht, the Netherlands

Notes

1. Governance describes a collaboration of public and/or private actors that aims to realize collective goals (Benz, 2005). A mode of governance is understood here as a particular arrangement of (public and/or private) actors embedded in certain institutions (context, legal framework), and a specific approach to the content (the object) (Driessen, Dieperink, Laerhoven, Runhaar, & Vermeulen, 2012).
2. Although some water infrastructures are also on the surface, the main infrastructural grid is considered as below surface.

References

Benz, A. (2005). Governance. In E.-H. Ritter (Ed.), *Handwörterbuch der Raumordnung* (4th ed., pp. 404–408). Hannover: ARL.

Billé, R. (2008). Integrated coastal zone management: Four entrenched illusions. *Surveys and Perspectives Integrating Environment and Society, 1*(2), 75–86. doi:10.5194/sapiens-1-75-2008

Biswas, A. K. (2004). Integrated water resources management: A reassessment. *Water International, 29*(2), 248–256. doi:10.1080/02508060408691775

Brown, J. D., & Damery, S. L. (2002). Managing flood risk in the UK: Towards an integration of social and technical perspectives. *Transactions of the Institute of British Geographers, 27*(4), 412–426. doi:10.1111/1475-5661.00063

Calder, I. R. (2005). *Blue revolution: Integrated land and water resources management* (2nd ed.). Earthscan: London [u.a.].

Driessen, P. P. J., Dieperink, C., Laerhoven, F., Runhaar, H. A. C., & Vermeulen, W. J. V. (2012). Towards a conceptual framework for the study of shifts in modes of environmental governance - experiences from The Netherlands. *Environmental Policy and Governance, 22*(3), 143–160. doi:10.1002/eet.1580

Dworak, T., & Görlach, B. (2005). Flood risk management in Europe @ the development of a common EU policy. *International Journal of River Basin Management, 3*(2), 97–103. doi:10.1080/15715124.2005.9635249

Dyckman, C. S., & Paulsen, K. (2012). Not in my watershed! Will increased federal supervision really bring better coordination between land use and water planning? *Journal of Planning Education and Research, 32*(1), 91–106. doi:10.1177/0739456X11426877

Edelenbos, J., Bressers, N., & Scholten, P. (2013a). Conclusions: Towards a synchronization perspective of connective capacity in water governance. In J. Edelenbos, N. Bressers, & P. Scholten (Eds.), *Water governance as connective capacity* (pp. 333–351). Burlington, VT: Ashgate.

Edelenbos, J., Bressers, N., & Scholten, P. (Eds.). (2013b). *Water governance as connective capacity.* Burlington, VT: Ashgate.

Gilissen, H., van Rijswick, M., & van Schoot, Td. (op. 2009). *Water en ruimte: De bescherming van watersysteembelangen in het ruimtelijk spoor* (1st ed.). *Ruimtelijk relevant: Vol. 1.* Amsterdam: Berghauser Pont Publishing.

Gleick, P. H. (2000). A look at twenty-first century water resources development. *Water International, 25*(1), 127–138. doi:10.1080/02508060008686804

Grigg, N. S. (2008). Integrated water resources management: Balancing views and improving practice. *Water International, 33*(3), 279–292. doi:10.1080/02508060802272820

Hartmann, T., & Albrecht, J. (2014). From flood protection to flood risk management: Condition-based and performance-based regulations in German water law. *Journal of Environmental Law, 26*(2), 243–268. doi:10.1093/jel/equ015

Hartmann, T., & Driessen, P. P. (2013). The flood risk management plan: Towards spatial water governance. *Journal of Flood Risk Management,* n/a. doi:10.1111/jfr3.12077

Hartmann, T., & Juepner, R. (2014). The Flood Risk Management Plan: An Essential Step Towards the Institutionalization of a Paradigm Shift. *International Journal of Water Governance, 2*(2), doi:10.7564/13-IJWG5

Hartmann, T., & Spit, T. (2012). Managing riverside property: Spatial water management in Germany from a Dutch perspective. In T. Hartmann & B. Needham (Eds.), *Planning by law and property rights reconsidered* (pp. 97–114). Farnham, Surrey: Ashgate.

Hartmann, T. (2011). *Clumsy floodplains: Responsive land policy for extreme floods.* Farnham, Surrey: Ashgate.

Heer, J. d., Nijwening, S., Vuyst, S. de, van Rijswick, M., Smit, T., & Groenendijk, J. (2004). *Towards integrated water legislation in The Netherlands: Lessons from other countries.* Retrieved from http://www.uu.nl/faculty/leg/NL/organisatie/departementen/departementrechtsgeleerdheid/organi-satie/onderdelen/centrumvooromgevingsrechtenbeleid/publicaties/Documents/TowardsIWL-final-report.pdf

Jønch-Clausen, T., & Fugl, J. (2001). Firming up the conceptual basis of Integrated Water Resources Management. *International Journal of Water Resources Development, 17*(4), 501–510. doi:10.1080/07900620120094055

Jong, P. (2007). The water system and water chain in Dutch water and environmental legislation. *Law, Environment and Development Journal, 2007*(2/3), 202–216. Retrieved from http://www. lead-journal.org/content/07202.pdf

Loucks, D. P. (2000). Sustainable water resources management. *Water International, 25*(1), 3–10. doi:10.1080/02508060008686793

Moss, T., & Monstadt, J. (2008). *Restoring floodplains in Europe: Policy contexts and project experiences.* London: IWA Publishing.

Moss, T. (2004). The governance of land use in river basins: Prospects for overcoming problems of institutional interplay with the EU Water Framework Directive. *Land Use Policy, 21*, 85–94. doi:10.1016/j.landusepol.2003.10.001

Moss, T. (2009). Zwischen Ökologisierung des Gewässerschutzes und Kommerzialisierung der Wasserwirtschaft: Neue Handlungsanforderungen an Raumplanung und Regionalpolitik. *Raumforschung und Raumordnung, 67*, 54–68. doi:10.1007/BF03183143

Perry, C. H., Miller, R. C., & Brooks, K. N. (2001). Impacts of short-rotation hybrid poplar plantations on regional water yield. *Forest Ecology and Management, 143*, 143–151. doi:10.1016/S0378-1127(00)00513-2

Scherer, D. (1990). *Upstream downstream: Issues in environmental ethics.* Philadelphia, PA: Temple University Press.

Wesselink, A. (2007). Flood safety in the Netherlands: The Dutch response to Hurricane Katrina. *Technology in Society, 29*(2), 239–247. doi:10.1016/j.techsoc.2007.01.010

Wiering, M., & Immink, I. (2006). When water management meets spatial planning: A policy-arrangements perspective. *Environment and Planning C: Government and Policy, 24*, 423–438. doi:10.1068/c0417j

Groundwater governance and spatial planning challenges: examining sustainability and participation on the ground

Gabriela Cuadrado-Quesada

Faculty of Law, University of New South Wales (UNSW), Sydney, Australia; Connected Waters Initiative Research Centre (UNSW), Sydney, Australia; National Centre for Groundwater Research and Training, Sydney, Australia

This article explores the role of participation and the conditions for achieving sustainable groundwater governance and spatial planning by empirically examining cases in Australia and Costa Rica. A comparison of the two indicates that crisis can increase the likelihood of participation; participation can influence the government to develop environmental legislation; self-interest and profit motivation can help promote sustainability under certain conditions; and environmental legislation tends to foster sustainability.

Introduction

Several current groundwater governance and spatial planning challenges are beset with high levels of complexity, conflict and uncertainty – so-called 'wicked' problems. These dilemmas are increasingly encountered in the public policy and academic spheres as groundwater resources become depleted or polluted. Along with groundwater resources, this paper focuses on vertical frontier problems, which are between groundwater and surface land use. Thus, the role of spatial planning for the discussion is fundamental.

In the last decade there has been a concerted effort to focus attention on groundwater governance internationally and domestically in developed and developing countries. For example, this trend is evident at the international and transnational level with the joint project supported by the Global Environmental Fund (GEF), Groundwater Governance: A Global Framework for Country Action; and in the European Union with the Water Framework Directive (Directive 2000/60/EC, European Parliament), and the Directive on the Protection of Groundwater against Pollution and Deterioration (Directive 2006/118/EC, European Parliament).

Australia has developed the intergovernmental agreement on the National Water Initiative (2004), which acknowledges the importance of groundwater and is committed to a whole water cycle approach. In developing countries similar endeavours have been proposed. In Costa Rica, the new Integrated Management Water Law (Exp. 17.742 Ley para la Gestión Integrada del Recurso Hídrico) was approved on 31 March 2014, but has not entered into force as it is under review at the Constitutional Chamber. This law recognizes the importance of the whole water cycle when dealing with water resources and promotes basic principles such as sustainable use and participation.

These developments seek to focus on groundwater resources and to reform its governance – filling gaps in legal frameworks, adapting traditional mechanisms applied to surface water and developing novel groundwater-specific arrangements. A common aim of many of these developments is to achieve sustainable and participatory outcomes.

There is a growing interest in the literature in examining how sustainability and participation interact with and guide practices of groundwater governance (Holley & Sinclair, 2013; Varady et al., 2012); and whether, through the implementation of them, more effective groundwater governance can be achieved (Godden & Peel, 2010). Despite this interest, it is still unclear how best to implement these conditions or principles, especially when dealing with spatial planning (Varady et al., 2012).

Therefore, this article explores the conditions under which participation can address the challenges of sustainable groundwater governance and spatial planning. It does so by conducting a comparative examination of two case studies and their efforts to achieve participatory and sustainable outcomes: Angas Bremer in South Australia, Australia; and Guácimo in the Caribbean in Costa Rica. Along the way, these cases provide insights for the vertical frontier governance framework. The specific findings suggest diverse approaches of how governmental and non-governmental stakeholders can achieve commitments within a specific governance arrangement.

Drawing on 30 interviews, the examination provides comparative insights into challenges faced by groundwater governance and spatial planning. Angas Bremer reveals insights into how a crisis unlocked the participation of the community which in the end took the shape of collaboration with the government to address the problems. In contrast, research in Guácimo reveals insights into how participation took the shape of social struggle to force the local government to put into force legislation to address the situation.

While each case throws up a range of different issues, comparison between them provides insights into two common and interrelated conditions, namely crisis unlocks participation and participation fosters sustainability. Nevertheless, the levels of participation and sustainability shown in the cases vary completely. The findings enable the article to suggest some important empirically based lessons for both groundwater policy-makers and groundwater governance and spatial planning literature.

Following this introduction, the paper is structured as follows. First it provides an overview of the methodology, followed by the two case studies. Then it turns to the discussion of the cases' efforts to address the challenges of groundwater governance and spatial planning. The final section takes stock of the findings to draw out conclusions and recommendations. The Appendix provides some key definitions.

Methods

This article follows a comparative case study approach to examine the phenomena of interest, namely groundwater governance and its relation with spatial planning. To select the cases, two steps were required: selecting a set of countries (Australia and Costa Rica); and selecting the case studies within the countries. In terms of selecting the countries, the following criteria were taken into account: different cultures and traditions in legal system (common law and civil law), different climatic conditions (semi-dry and tropical), developed and emerging countries and democratic countries that respect the rule of law.

For the case studies within the countries, a number of groundwater governance experiences were reviewed to determine whether at least some of their components embraced widely recognized characteristics of participation and sustainability. Another consideration for selecting the cases was to capture a diversity of conditions, including cases that focused on different problems and diverse solutions.

The research relied primarily on qualitative interviewing (Liamputtong & Douglas, 2005) with some doctrinal analysis. The interview selection process was based on purposive sampling (Patton, 1990), selecting interviewees to represent key stakeholder groups involved in groundwater governance. A total of 30 interviews were conducted: 15 in Angas Bremer and 15 in Guácimo. All interviews were in-depth conversations and followed a semi-structure interviewing technique.

Data analysis followed the stipulations of adaptive theory which offers an iterative strategy of enquiry that uses pre-existing theory to guide study, while simultaneously generating theory from data analysis in the conduct of empirical research (Layder, 1998). The validity of conclusions was heightened by triangulating multiple resources of data, i.e. interviews, literature and legislation, and a process of respondent evaluation conducted near the end of the fieldwork (Angrosino & Mays de Perez, 2003).

Case studies

This section begins with an explanation for selecting the cases; then it provides an overview of the legislation in each country; and finally it examines in detail the case studies. It analyses how participation was developed to address groundwater and spatial planning challenges and how sustainable groundwater governance and spatial planning can be achieved. As will be shown, there are similarities and variations across the cases regarding both the nature of participation, the approach to sustainability and the legal/ voluntary mechanisms implemented.

Angas Bremer in South Australia, Australia

This experience was chosen because the Angas Bremer District, located in the state of South Australia, is an area of high groundwater use, but with recognized innovation by the community in collaborative groundwater governance (Thompson, 2008). The district is located near the town of Strathalbyn, beside Lake Alexandrina, and 60 km south-east of Adelaide, South Australia. The district is named after the Angas and the Bremer rivers. Currently, it is one of Australia's premium wine-grape regions. Due to the growth of several crops (e.g. lucerne and potatoes) and farming, some environmental problems have raised, such as overuse of groundwater and salinity. All these conditions led this region to face a water crisis.

Legislation

Australia has included the principle of sustainability in national laws and policies (Lyster, Lipman, Franklin, Wiffen, & Pearson 2012); examples of this are the Australia's National Strategy for Ecologically Sustainable Development of 1992, and the National Water Initiative, which states that 'governments have a responsibility to ensure that water is allocated and used to achieve socially and economically beneficial outcomes in a manner that is environmentally sustainable' (National Water Initiative, 2004, preamble, para. 2). In addition, there have been environmental lawsuits substantiated in this principle.[1] At the same time, Australia has recognized the importance of the participation principle in environmental issues. Well-known examples of this in Australia are the regional natural resources management initiatives

(Holley, 2010). Australia has recognized, in particular, the importance of the participation principle under the National Water Initiative (National Water Initiative, 2004, para. 36).

In Australia, planning and groundwater regimes remain distinct. Nevertheless there have been some attempts to integrate planning to water regimes, i.e. requiring consistency between plans. The most prominent legal document for water regulation is the Intergovernmental Agreement on the National Water Initiative. This is a national instrument that has been adopted by all the states in Australia. Australia's National Water Initiative and its emphasis on sustainable groundwater governance (Plazinska, 2007) is considered by many to be a world leader in water governance reform (Godden & Foerster, 2011).

The primary regulatory authority for water resources management is vested in the state minister who is responsible for the administration of the water resources legislation. This is done at the state and territory level. Each state jurisdiction provides for some form of consultation by the regulatory agencies with advisory bodies in the governance of water resources, especially with respect to water planning (Gardner, Bartlett, & Gray, 2009).

A prominent example of this type of body in Australia is in South Australia (SA). The SA Act established the Natural Resources Management (NRM) Council as a standing state-level advisory body. In order to assist with the performance of its functions, the NRM Council has the power to appoint committees and delegate functions and powers (Gardner et al., 2009).

In South Australia it is the regional Natural Resource Management (NRM) board, established by the minister, that performs subsidiary functions under the Natural Resources Management Act 2004 (SA). These boards are established to undertake an active role with respect to the management of natural resources within its region. A regional NRM board must work with local committees. Such a committee has been the Angas Bremer committee.

Overview of the case

In Angas Bremer district, by 1981 the annual use of groundwater for irrigated agriculture had increased to four times the annual groundwater recharge. The annual groundwater use had reached 26,600 megalitres, four times the estimated annual recharge of 6000 mega-litres (Thomson, 2002). This makes clear the unsustainable use of groundwater resources. In order to resolve the abovementioned problems, the irrigators and the community in general had to organize themselves (they formed the Angas Bremer committee) and have implemented different measures. The role of the government has been vital for the success of this experience. For example, the over-used groundwater resources were brought under the control of legislation, but also they have implemented voluntary initiatives.

One of the first voluntary initiatives taken by the irrigators was to change their crops to more water-efficient ones. For example, many of them changed from lucerne to grapes. As one irrigator expressed:

> We have changed our crops, we went out of irrigated lucerne and virtually none grown now, we're using more water efficient crops, mainly vineyards. [...] Lucerne I think went from something like 6 or 7 thousand hectares to zero during that time, predominantly driven by several reduced water use. (Irrigator 6, personal communication, 12 June 2013)

This shows the concern of the community in reducing water use. Another reason for many growers to change to grapes crops was an economic one – grapes are more profitable than lucerne.

Moreover, a land-use survey was performed to document the area of each crop type that was being grown by each irrigator (Thomson, 2002). As one irrigator explained: 'We were allocated a volume of water equal to the sum of the theoretical amounts needed to irrigate the area of each of our crops but we were using less' (Irrigator 1, personal communication, 13 June 2013). This means that even though the irrigators were having a significant water allocation, they were using much less because they were convinced that the total volume allocated greatly exceeded the volume thought to be sustainable.

As another interviewee mentioned: 'By working together we have developed and implemented different innovative water, groundwater and land management policies' (Governmental agency 9, personal communication, 14 June 2013). A condition that enabled everybody to work together was the understanding that they were facing a crisis. As argued by an interviewee:

> Everybody could see there was a problem, the department became involved and obviously we had to take less water, we didn't know how the system worked. I think the nutshell of that was that the community saw for itself there was a crisis. (Local resident 2, personal communication, 15 June 2013)

Other initiatives include:

> the exchange of groundwater licenses for lakewater licenses, the aquifer storage and recovery (it is achieved by directing winter floodwater into the bores), building locally-funded pipe-lines, changing the land use, doing re-vegetation; and developing new tools based on scientific data – such as the Commonwealth Scientific and Industrial Research Organisation, (CSIRO) FullStop device, which detects whether the soil is wetted and measures the amount of salt, or other chemicals. (Local resident 3, personal communication, 13 June 2013)

Additionally, some legal initiatives have been implemented, including new requirements on water licences – such as the irrigation annual reporting. These initiatives were taken because the government has been working with the committee and the community, and has been providing funding to do the monitoring. As commented by a participant:

> Another departmental person was on-board real early with us and developed a lot of water management practices, water conservation, measurements, and annual monitoring came in. We got funding from the government to do the monitoring. So it was a really successful period, totally in my mind put down to strong community participation in partnership with the government. (Local resident 11, personal communication, 14 June 2013)

The irrigators, the committee and the community have been very effective at addressing some spatial planning and groundwater challenges. For example, changing land-use patterns, i.e. re-vegetation, as well as achieving more sustainable water use. As argued by an interviewee:

> That team worked brilliantly, there was a lot of argy-bargy and a lot of hard times and a lot of arguments but in the end what the government was able to produce was a water management plan that the community didn't disagree with. The implication was, we cut our groundwater allocation. I think we reduced our water allocation by a third voluntarily. (Irrigator 2, personal communication, 13 June 2013)

Keys to this success have included strong participation involving the community, irrigators and experts from government agencies. A condition that facilitated this has been proactive community leadership. As expressed by one interviewee:

> Over a couple of years then there was a water management plan and the Angas Bremer committee was formed. We had a good high level departmental people on a committee with proactive community leaders and representatives of the community, so it was really good. (Local resident 8, personal communication, 15 June 2013)

Summarizing, this case study demonstrates some conditions that contribute to participation to address the challenges of groundwater governance and spatial planning which are as follows: crisis which includes groundwater over-use and salinity, proactive community leaders like the founding members of the Angas Bremer committee, and support from governmental agencies. Additionally, it illustrates some conditions that foster sustainable groundwater and spatial planning, such as self-interest or profit motivation and legislation.

Even though the experience of Angas Bremer has been successful, the district is facing noteworthy challenges. As commented by an interviewee, 'Keeping people involved and actively participating in the committee, especially young people has been really difficult' (Local resident 3, personal communication, 14 June 2013). Similarly another participant expressed:

> Farmers or people that are there [...] are just there off their own back trying to do something positive towards the local irrigators system. It is hard to get people on there; no one wants to be on there really. I think because things are back to smooth sailing again, through the drought we were busy, not busy but there was more pressure on us. (Local resident 6, personal communication, 13 June 2013)

Additionally, there is a problem with the limits of volunteerism, i.e. lack of money and time. As argued by a participant:

> I think you could summarise the committee's activities now as fighting for their existence to keep the door open. [...] They spend most of the time applying for grants to keep the door open, paper shuffling, most of the community's lost interest because there's no results. (Local resident 7, personal communication, 15 June 2013)

Thus, participation becomes vulnerable to falling over when government might remove its support.

Guácimo, Caribbean of Costa Rica

In Costa Rica, Guácimo of Limón, Caribbean, was chosen to represent problems and solutions, done by the communities and local governments, due to groundwater quality issues arising from growing agricultural plantations. Guácimo has been affected by growing agricultural plantations, especially banana and, more recently, pineapple. Nowadays, the pineapple plantations present a major problem in this region due to the pollution of groundwater by agrochemicals used in pineapple plantations (Ruepert, 2005).

Legislation

Costa Rica has also recognized the importance of the principle of sustainability and has incorporated this principle in national laws; an example of this is article 2 of the Environmental Law (Ley Orgánica del Ambiente, 1995), and article 2 d of the Integrated Management Water Law (2014). As well, this principle has substantiated several environmental lawsuits. One example of this is the Resolution of the Constitutional Chamber 5689–08 where it was recognized that the pineapple plantations are violating sustainable development and they should stop immediately the activities that are putting sustainable development in danger. Additionally, Costa Rica has recognized the significance of the principle of participation in environmental issues. A prominent example for Costa Rica is article 6 of the Environmental Law (1995) and article 2 g of the Integrated Management Water Law (2014).

In Costa Rica the governing body for water resources is vested in the Ministry of Environment and Energy. It is responsible for defining the policies and strategies that concern the sector. Nevertheless, there are several institutions that in one way or another also regulate the water sector. There is the Costa Rican Institute of Aqueducts and Sewage (AyA), which is the lead agency with regard to the supply of water and sewage systems. Closely linked to the AyA are the Administrative Associations of Communal Aqueducts (ASADAS), which are associations managed by the community responsible for the water supply in their community. Then there is the National System of Groundwater, Irrigation and Drainage (SENARA), which is the body responsible for the capabilities of the groundwater basins and the margins of groundwater exploitation in the country. Finally, there is the Ministry of Health, which is the institution responsible for guaranteeing the availability and quality of water. Therefore, sometimes there is strong overlap of responsibilities and an important institutional confusion (Astorga, 2009).

Another problem that Costa Rica faces is the lack of funding to perform hydrogeological studies, which has been a determinant factor in increasing groundwater governance challenges (Arias, 2009). At the same time a real quantification of groundwater resources does not exist, nor an integrated plan of groundwater resources (Astorga, 2009). Groundwater governance is not only messy but also based on demand and not on the existing availability of groundwater resources, because they are unknown (Arias, 2009).

Overview of the case study

Costa Rica has converted itself into one of the world's leading pineapple producers, and the world's number one exporter. With more than 200% growth compared with the year 2000, official numbers estimate that there are approximately 50,000 hectares dedicated to this crop (Cuadrado-Quesada, 2008), which is exported primarily to Europe and the United States. Most of the pineapple companies are subsidiaries of transnational companies such as Del Monte and Chiquita. Some of the main problems arising from the pineapple plantations include: the change in land use of an important number of hectares previously designated to protect the forest and water table areas; invasion of protected areas around rivers and springs; and groundwater pollution by an indiscriminate use of agrochemicals (Cuadrado-Quesada, 2008). The first published study saying that the groundwater was polluted was released in 2005 (Ruepert, 2005).

Due to all these problems, the communities of Guácimo have actively participated in addressing groundwater governance and spatial planning problems. The participation in this case study has been very different from the participation previously discussed in the

Angas Bremer case. The 'shapes' of participation taken in this case include demonstrations and lawsuits. Responses to the actions taken by the communities of Guácimo have had different solutions and scope from those taken in the case study of Angas Bremer. However, like the Angas Bremer case, an important condition that enabled people from the communities to start organizing themselves was the understanding that they were suffering a crisis. As argued by an interviewee, 'We knew that there were many problems, we saw the rivers polluted, we saw the erosion, the dead fish [...] all that happened soon after the pineapple plantation came to our community' (Local resident 26, personal communication, 15 June 2013).

As previously mentioned, in this case study the 'shape' of participation taken by the communities of Guácimo has been of *social struggle*. As pointed out by an interviewee, 'We, the communities, feel that the governmental institutions don't care about us, the animals, or the natural resources, they just continue to favor pineapple companies no matter what they do, that is why we have organized several demonstrations' (Local resident 29, personal communication, 13 June 2013). Under the pressure of people who were demonstrating in front of the municipality and participating in all public meeting where issues concerning pineapple plantations were discussed, the municipality has established a moratorium since 2008 to control the future expansion of pineapple plantations. This measure was taken as part of the spatial planning in the canton. The objective of the moratorium was to stop the pineapple plantations in the south of the canton, where all the recharge areas of the aquifers are located (Concejo Municipal de Guácimo, 2008). This legal tool helped to stop further groundwater pollution and protect fundamental spots such as the recharge areas of the aquifers.

The municipality had this moratorium for approximately five years. As argued by an interviewee:

> We knew we were facing a crisis. We felt that the Municipality had to do something, people were angry, they wanted actions, and we already knew that other communities of Limón were having their water polluted by agrochemicals used in pineapple plantations. (Government official 20, personal communication, 12 July 2013)

Additionally to the pressure measures taken against the municipality, the community leaders formed an organization to protect natural resources. This organization is called the Environmental Association Pro-defense of the Natural Resources of the Caribbean. As discussed by a participant, 'Among the achievements of our association was stopping a new pineapple plantation in the community of Iroquois, which we knew it will invade the forest, the protected areas and the recharge areas; and eventually it will pollute the water' (Community Based Organisation (CBO) 25, personal communication, 14 July 2013).

In this case, the community and the Environmental Association Pro-defense of the Natural Resources of the Caribbean presented a lawsuit against the pineapple plantation. As argued by an interviewee:

> The National Technical Environmental Secretary (SETENA) resolved not to give the environmental viability, which in accordance with the Environmental Law, is needed to start any significant activity or project. In the end, this governmental agency didn't give the environmental viability to the pineapple plantation, thus, they couldn't start with the project. (CBO 28, personal communication, 15 July 2013)

This case study demonstrates some successes in addressing the challenges of groundwater governance and spatial planning. The conditions that facilitated these successes are as follows: crisis due to the expansion of the pineapple plantations and groundwater pollution, proactive community leaders (e.g. the founding members of the Environmental Association Pro-defense of the Natural Resources of the Caribbean), and involvement of the municipality. Additionally, it illustrates some conditions that foster sustainable groundwater and spatial planning such as legislation/moratorium.

Even though the moratorium was an essential precautionary legal tool, part of the spatial planning in Guácimo, and indeed helped the canton to protect its aquifers, it is no longer in force. The National Chamber of Pineapple Producers and Exporters (Cámara Nacional de Productores y Exportadores de Piña – CANAPEP) successfully presented a legal action (Amparo) against the municipality, arguing that the municipality was violating their right to free trade and private property. They won the case on 18 October 2013. As commented by an interviewee, 'Keeping the moratorium was fundamental for the protection of our aquifers. It is obvious that the Constitutional Court doesn't care about the communities or water resources. They just care about trade' (Local resident 30, personal communication, 26 June 2014).

Furthermore, there is a problem with the limits of volunteerism, i.e. lack of money and time. As discussed by a participant 'We are getting tired of demonstrating and suing the companies and the governmental institutions, some people have lost interest because there's no results' (Local resident 27, personal communication, 15 June 2013). Thus, participation becomes likely to falter when people start feeling that their efforts do not result in improvements.

Discussion and comparison of the cases

Numerous conditions under which participation can address the challenges of groundwater governance and spatial planning were revealed in the finer detail in both case studies. Aforementioned conditions are crisis, proactive community leaders and active involvement of governmental agencies.

Regarding crisis, even though the Angas Bremer and Guácimo cases were facing a crisis, the type and scope were very different. The Angas Bremer case was facing a quantity problem due to overuse; the Guácimo case was dealing with a quality issue. Even more important, the Guácimo case was struggling with pollution due to an economic activity that in principle should have been properly regulated by governmental institutions. Nevertheless, the result was significantly similar: the crisis unlocked citizen participation.

Concerning proactive community leaders, in their own way, both cases, Angas Bremer and Guácimo, appeared to have good and influential leaders. In the Angas Bremer case, the members of the committee who were mainly irrigators helped first to reduce water use, then to foster participation and finally to embrace local collective management, as discussed by a participant. Due to this local collective management, it has been possible to develop and implement different innovative groundwater and spatial planning policies such as monitoring and re-vegetation (Governmental agency 27, personal communication 15 July 2013). In the Guácimo experience, it was evident how the leaders in the communities organized themselves and created a CBO in order to establish lawsuits against pineapple plantations. Additionally, they pressured local government to get involved and take legal actions.

With regard to the involvement of governmental agencies, in both cases the importance of working with these agencies led demonstrably to better outcomes. Working together the communities and government successfully addressed groundwater governance and spatial planning challenges. In the case of Guácimo, the role of the municipality has been outstanding, as discussed by a participant. Putting into force a spatial planning regulation such as the moratorium and with that stopping the expansion of more pineapple plantations in the recharge area of the aquifers has been an effective measure to address groundwater governance and spatial planning challenges; nevertheless, it is no longer in force (Local resident, 29, personal communication, 16 July 2013).

In the Angas Bremer case, the role of the NRM Board and the Department of Environment, Water and Natural Resources has been fundamental for the successes of the committee and the community. The Angas Bremer committee has had economic support from the NRM Board through grants. Additionally, it has received grants from the South Australia state government. Nevertheless, the committee and the whole community are now facing a lack of funding from governmental agencies. Thus, the withdrawal of funding suggests that participation becomes vulnerable to falling over. At the same time, some conditions under which sustainable groundwater governance and spatial planning can be achieved were demonstrated in much detail in both case studies. These conditions are legislation and self-motivation/profit motivation.

Regarding legislation, it was shown that in the case study of Guápiles the municipality established a moratorium for future expansion of pineapple plantations. This moratorium was implemented because of people's pressure towards the municipality. In addition, legislation also helped to achieve a bit more sustainable groundwater governance and spatial planning when the Environmental Association Pro-defense of the Natural Resources of the Caribbean sued a pineapple plantation that wanted to start operations in an aquifer recharge area in the community of Iroquois.

In the Angas Bremer case study, legislation has been an important tool to achieve more sustainable groundwater governance and spatial planning. An example of this has been implemented in the water licence requirements, for example, with the monitoring of groundwater use, which helps to use water in a more sustainable way. Nevertheless, the community has gone further. Besides complying with legal requirements, it has implemented many more innovative and voluntary mechanisms such as the exchange of groundwater licences for lake water licences, aquifer storage and recovery, locally funded pipelines, re-vegetation; and using the FullStop device.

Concerning self-motivation/profit motivation, it was illustrated with the Angas Bremer case study that voluntary initiatives such as changing crops to more water-efficient ones are easily implemented when accompanied with a profit motivation. Vineyards are more profitable than crops like lucerne or potatoes. In the case study of Guápiles, the self-motivation has been the fear of having the groundwater polluted with agrochemicals. The people in Guácimo know that many neighbouring communities such as Milano, Cairo, Luisiana and Francia have already their water polluted with agrochemicals; therefore, they have tried to avoid this happening in their communities.

Conclusions and recommendations

This article has investigated the conditions under which participation can address the challenges of groundwater governance and spatial planning; and the conditions under which sustainable groundwater and spatial planning can be achieved in very different scenarios. Drawing on 'on-ground' experiences, it has provided insights into a number of significant 'gaps' in groundwater and spatial planning policy and literature.

The conditions under which participation can address the challenges of groundwater governance and spatial planning identified were revealed in both case studies. These conditions are as follows:

- Crisis: environmental/water, quantity and quality.
- Proactive community leaders: organized in CBOs and committees.
- Involvement of governmental agencies: interested government employees and political will.

More interestingly, the finding in these two cases was the different approaches that participation has taken in each case study. The participation in the Angas Bremer case study is closely related to collaboration, where government and community worked together and where the government facilitates the work done by the citizens. The shape of participation observed in the Guápiles case study is public participation 'pressuring' the government to take legal actions. Nevertheless, in the end the role of the municipality was efficient. The municipality of Guácimo is the only municipality in Costa Rica that has implemented a moratorium to regulate the expansion of pineapple plantations.

In a similar manner, some conditions under which sustainable groundwater governance and spatial planning can be achieved were shown in the case studies in Angas Bremer and Guácimo. These conditions are as follows:

- Legislation: water licence requirement, i.e. monitoring of groundwater use, and spatial planning regulation, i.e. moratorium.
- Self-motivation/profit motivation: more water-efficient but also more profitable crops and access to safe drinking water.

Moreover, the findings indicate that sustainability has been also understood differently in both cases. In Angas Bremer, it includes a more integrated approach that is motivated by self-motivation or profit motivation such as changing to more water-efficient crops, but it then helped to make sustainability more holistic. In Angas Bremer, the community reduced groundwater use and implemented monitoring, but at the same time it went further: they exchanged groundwater licences for lake water licences; they implemented aquifer storage and recovery; and they established re-vegetation areas, among others. In Guápiles, the approach has been building some bases to stop pollution through legislation, i.e. the moratorium. Regrettably this legal tool in no longer in force; therefore, the municipality is facing the challenge to implement new legal and policy instruments to offer the communities a sustainable development and effective water protection.

For the vertical frontier framework, these specific findings suggest diverse approaches of how governmental and non-governmental stakeholders can achieve some commitments within a certain governance arrangement. Moreover, the Guácimo experience shows how public participation, as defined by Arnstein (1969), can induce, to some level, important environmental reforms. In this specific instance public participation took the form of social struggle.

Additionally, these two case studies demonstrate how groundwater governance is directly interconnected to spatial planning regulations such as land use as illustrated by Foster, Garduño, Tuinhof & Tovey (2010). Therefore, it is important to take this interconnection into consideration when discussing and implementing policies and laws. As important as the interconnection between groundwater governance and spatial planning is the fundamental role that public participation and sustainable use have when designing and formulating laws and policies because without acknowledging these principles there cannot be effective groundwater governance.

What may be needed, then, are more proactive conditions for addressing the challenges of groundwater governance and spatial planning. In the case of Costa Rica, this would clearly focus on having local governments and national governmental agencies, e.g. the Constitutional Chamber worrying more about sustainable groundwater governance and sustainable spatial planning mechanisms instead of focusing solely on economic development. The municipality of Guácimo has done an important job in trying to address this issue. Unfortunately, the Constitutional Chamber considered that the rights to free trade and private property are more important than precautionary measures to protect groundwater resources.

Another recommendation for the Australia case study is obviously to keep the state government developing greater funding and support matched to the task at hand, which is vital to reducing cost to volunteers and ensuring a more effective organization and implementation process. In this context, what is needed is to explore how to create institutional mechanisms and ways of redirecting funding to provide more sustained community organizations.

A further recommendation for the two case studies is to foster networks and sharing experience among these kinds of groups. For example, what the Angas Bremer committee and community have achieved will be worthwhile for other committees in South Australia and around the whole country. In the case study from Costa Rica it would be worthwhile to share the experience on how they implemented the moratorium (and how they kept it in force for approximately five years) with all the other cantons where there are pineapple plantations, so this legal mechanism can also be implemented, at least as a temporary measure.

Acknowledgements

The author would like to thank Dr Cameron Holley and the anonymous reviewers for valuable comments and suggestions. Additionally, all interviewees are thanked for volunteering to participate in this research.

Funding

This research was funded by the Faculty of Law, University of New South Wales (UNSW), the Connected Waters Initiative Research Centre, and the National Centre for Groundwater Research and Training (NCGRT).

Note

1. One example of this is the case of *Bentley v BGP Properties Pty Limited*. The judge stated that the conservation of biological diversity and ecological integrity is one of the pillars of ecologically sustainable development (*Bentley v BGP Properties Pty Limited*, 2006).

References

Acta [Act] No 34–2007, Sesión Ordinaria [Ordinary Session] No 27–2007, celebrada por el Concejo Municipal de Guácimo [Municipal Parlament], el 3 de Julio del 2007. Publicada en la Gaceta No 126 del 1 de Julio del 2008.

Angrosino, M., & Mays de Perez, K. (2003). Rethinking observation: From method to context. In N. Denzin & Y. Lincoln (Eds.), *Strategies of qualitative inquiry* (2nd ed.). Thousand Oaks, CA: Sage Publications.

Arias, M. (2009). *Gestión del Recurso Hídrico y Uso del Agua* [Water management and water use]. San José, Costa Rica: Decimoquinto Informe Estado de la Nación en Desarrollo Humano Sostenible/Programa Estado de la Nación.

Arnstein, S. R. (1969, July). A ladder of citizen participation. *JAIP, 35*(4), 216–224.

Astorga, Y. (2009). *La Gestión de Aguas Subterráneas: un espacio de conflicto* [Groundwater management: An area of conflict]. San José, Costa Rica: Programa Estado de la Nación en Desarrollo Humano Sostenible.

Case of *Bentley v BGP Properties Pty Ltd*. [2006] NSWLEC 34; (2006) 145 LGERA 234. Retrieved from http://www.lec.lawlink.nsw.gov.au/lec/issues_in_focus/biodiversity_cases.html. Land and Environment Court Biodiversity cases. Cases considering ecologically sustainable development – precautionary principle.

Cuadrado-Quesada, G. (2008). Legalización de la contaminación de las agua para consumo humano [Legalising the pollution of water for human consumption] (caso del diurón y el bromacil), Ambietico No 177, Heredia, Costa Rica.

Cuadrado-Quesada G., & Castro R. (Eds.). (2008). *Protegiendo hoy el agua del mañana: experiencias comunitarias exitosas* [Protecting today the water for tomorrow: Successful community based experiences]. San José, Costa Rica: Environmental and Natural Resources Law Centre (CEDARENA).

Cullingworth, B., & Nadin, V. (2006). *Town and country planning in the UK* (4th ed.). London: Routledge.

Directive 2000/60/EC of the European Parliament of the Council of 23 October 2000 establishing a framework for Community action in the field of water policy. Available at: http://eurlex.europa.eu/LexUriServ/LexUriServ.do?uri=OJ:L:2000:327:0001:0072:EN:PDF

Directive 2006/118/EC of the European Parliament of the Council of 12 December 2006 on the protection of groundwater against pollution and deterioration. Available at: http://eurlex.europa.eu/LexUriServ/LexUriServ.do?uri=OJ:L:2006:372:0019:0031:EN:PDF

Exp.17.742 Ley para la Gestión Integrada del Recurso Hídrico [Integrated Management Water Law]. (2014). Costa Rica Asamblea Legislativa.

Foster, S., Garduño, H., Tuinhof, A., & Tovey, C. (2010). *Groundwater Governance, conceptual framework for assessment of provisions and needs*. Washington: Sustainable Groundwater Management, Contribution to Policy Promotion. The World Bank, Global Water Partnership and Water Partnership Program.

Gardner, A., Bartlett, R., & Gray, J. (2009). *Water resources law*. Australia: Lexis Nexus, Butterworths.

Gleick, P. H. (1998). Water in crisis: Paths to sustainable water use. *Ecological Applications, 8*(3), 571–579. doi:10.1890/1051-0761(1998)008[0571:WICPTS]2.0.CO;2

Godden, L., & Foerster, A. (2011). Introduction: Institutional transitions and water law governance. *The Journal of Water Law, 22*, 53–57.

Godden, L., & Peel, J. (2010). *Environmental law, scientific, policy and regulatory dimensions*. South Melboure, Victoria: Oxford University Press.

Hartmann, T., & Sipt, T. (2014). Editorial: Frontiers of land and water governance in urbanregions-Questions of sectors, scale and shifts of working paradigms. In prep.

Holley, C., & Sinclair, D. (2013). Deliberative participation, environmental law and collaborative governance: Insights from surface and groundwater studies. *Environmental and Planning Law Journal, 30*(1), 32–55.

Holley, C. (2010). Public participation, environmental law and new governance: Lessons from designing inclusive and representative participatory processes. *Environmental and Planning Law Journal, 27*, 360.

Knüppe, K., & Pahl-Wostl, C. (2011). A framework for the analysis of governance structures applying to groundwater resources and the requirements for the sustainable management of associated ecosystem services. *Water Resources Management, 25*(13), 3387–3411. doi:10.1007/s11269-011-9861-7

Layder, D. (1998). *Sociological practice linking theory and social research*. London: Sage.

Ley Orgánica del Ambiente [Environmental Law] Nº 7554 del 13 de diciembre de 1995.

Liamputtong, P., & Douglas, E. (2005). *Qualitative research methods* (2nd ed.). Melbourne: Oxford University Press.

Lyster, R., Lipman, Z., Franklin, N., Wiffen, G., and Pearson, L. (2012). *Environmental and planning law in NSW* (3rd ed.). Sydney: The Federation Press.

National Water Initiative. (2004). Retrieved from http://www.nwc.gov.au/__data/assets/pdf_file/0008/24749/Intergovernmental-Agreement-on-a-national-water-initiative.pdf

Patton, M. (1990). *Qualitative evaluation and research methods* (2nd ed.). Beverly Hills, CA: Sage Publications.

Plazinska, A. (2007). *Understanding Groundwater, Science for Decision Makers*, September 2007, Bureau of Rural Science, Department of Agriculture, Fisheries and Forestry, Australian Government.

Resolución de la Sala Constitucional No 5689–08, Costa Rica, 2008. [Resolution of the Constitutional Chamber]

Ruepert, C. (2005). *Groundwater vulnerability to pesticide contamination in Costa Rica*. Heredia, Costa Rica: IRET.

Sands, P. (2003). *Principles of international environmental law* (2nd ed.). Cambridge: Cambridge University Press.

Thompson, T. (2008). Water innovations in the Angas Bremer district of SA. *Water Down Under,* 14 April, Adelaide.

Thomson, T. (2002). Angas Bremer Irrigators lead the nation. *Article in Australian Viticulture, 6*(3), 51–53. ISSN 1329–0436.

United Nations Economic Commission for Europe. (2008). *Spatial planning, key instrument for development and effective governance with special reference to countries in transition*. New York and Geneva: UNECE.

United Nations Environment Programme. (2003). *Groundwater and its susceptibility to degradation: A global assessment of the problem and options for management*. Early Warning and Assessment Report Series, RS. 03–3. UNEP: Nairobi, Kenya.

Varady, R., Van Weert, F., Megdal, S., Gerlak, A., Abdalla-Iskandar, C., & House-Peter, L. (2012). *Groundwater policy & governance*. Rome: GEF.

Appendix: Definitions

Groundwater governance and spatial planning

Excluding water locked in polar ice caps, groundwater constitutes approximately 97% of all fresh-water potentially available for human use (United Nations Environment Programme (UNEP), 2003). Within this context, the contribution of access to groundwater is vital.

While groundwater use continues to increase, the governance of this natural resource remains a significant challenge for law and policy-makers. The international joint project supported by the Global Environmental Fund (GEF), Groundwater Governance: A Global Framework for Country Action, defines groundwater governance as:

> the process by which groundwater resources are managed through the application of responsibility, participation, information availability, transparency, custom and rule of law. It is the art of coordinating administrative actions and decision making between and among different jurisdictional levels – one of which may be global.

For this study, particularly relevant challenges include the interaction between groundwater governance and spatial planning. Spatial planning is defined as a coordination or integration of the spatial dimension of sectoral policies through a territorially based strategy (Cullingworth & Nadin, 2006). It addresses the tensions and contradictions among sectoral policies (United Nations Economic Commission for Europe (UNECE), 2008), e.g. for conflicts between economic development and groundwater protection. Spatial planning is critical for promoting sustainable use of land and natural resources.

Groundwater is highly dependent upon spatial planning regulations over land use (and changes in land use), especially in the main aquifer recharge areas (Foster, Garduño, Tuinhof, & Tovey, 2010). Provisions on groundwater are contained in legislation relating to spatial planning, e.g. the Costa Rican Forestry Law (Ley Forestal) establishes that the surrounding areas distance of 100 m from springs must be protected. Nevertheless, such legal instruments have been poorly implemented (UNECE, 2008).

In urban environments, land-use classification and control are generally the domain of the municipal or local government, and the absence of mechanisms whereby water resource agencies can influence the process is a frequent governance weakness (Foster et al., 2010). Moreover, in many developing countries such as Costa Rica, legislation to cope with undesirable land-use practices is often weakly enforced or even non-exist and progress with implementing controls in the interest of groundwater is highly dependent upon stakeholders awareness and participation (Knüppe & Pahl-Wostl, 2011).

Rural land-use practices and the intensification of agriculture production are strongly influenced by national agriculture and food policy (Foster et al., 2010) that trumps land-use planning, risking environmental damage (Ruepert, 2005). Although there are many principles or conditions that facilitate solving the challenges of spatial planning and water governance regimes, especially important are sustainability and participation.

Sustainability

The general norm in international law underlying the effort to achieve sustainable development is that of sustainability, a norm based on other norms such as respect for human life, for nature and its flora and fauna, for justice and development (Sands, 2003). Sustainability as a principle of international law is no more abstract, or more general, than other important norms such as peace, security and respect for human rights.

For the purpose of this research, sustainable groundwater use is when withdrawals do not exceed their own recharge rates (Gleick, 1998). It requires a system that enjoys protection measures that endure over time (Cuadrado-Quesada & Castro, 2008), e.g. protection areas. Finally, these protection measures are established by law, which will guarantee to some extent its enforcement through time. Sustainable spatial planning is understood as the implementation of the coordination and integration of the spatial dimension of sectoral policies through a sustainable territorially based strategy.

Despite the imminent importance of achieving sustainable use of groundwater resources through spatial planning, recent studies keep demonstrating the unsustainability of current groundwater resource use (Varady et al., 2012). This study focuses on examining under what conditions sustainable groundwater use and sustainable spatial planning can be achieved.

Participation

Public participation is the means by which those excluded from the political and economic processes can induce significant social reform which enables them to share in the benefits of more affluent members of society (Arnstein, 1969). Participation for the purpose of this paper is understood as a degree of power or control which guarantees that people can govern. This includes outright control/ governance and collaborative approaches. In particular, this study focuses in the conditions that assure that people can govern groundwater resources and spatial planning.

Governance frontier

The topic of this article lies on the *governance frontier* developed by Hartmann and Sipt (2014) that consists of three frontiers of land and water governance which emerge along the physical boundaries between land and water. In particular, this research has been developed within the vertical frontier with a special focus on groundwater and land use above. Often the interactions between land uses on top and water below – such as pollution or depletion of groundwater – occur quite unnoticeably. The surface is the boundary between water below and land on top. However, socio-economic and environmental changes challenge this boundary in various ways. This paper explores some of these challenges.

Impact of short-rotation coppice on water and land resources

Jens Hartwich, Jens Bölscher and Achim Schulte

Institute of Geographical Science, Applied Geography – Environmental Hydrology and Resource Management, Freie Universität Berlin, Berlin, Germany

The European Union is focusing on increasing renewable energy sources. One of these sources, known as short-rotation coppice (SRC), involves planting wood, as an energy carrier, on agricultural sites. By presenting a literature research, this paper studies the advantages and disadvantages of SRC in relation to its effects on water and land resources. In terms of renewable energy sources, considering these effects in the current process of social reconstruction is essential for sustainable development. With regard to this, SRC is a key element in the environmental management of land and water.

Introduction

In 2007 the European Council mandated that the proportion of renewable energy in electricity production should increase to 20% in the European Union by 2020 to reduce CO_2 emissions and lessen the effects of climate change. In Germany, after an intense public debate, the targeted percentage has been established at a minimum of 35%, out of which 8% should be met by bioenergy (and 27% from other renewable energy sources) (BMELV & BMU, 2010; BMU, 2012). Already, 92% of the heat production in Germany that is generated from renewable energy is gained from biomass, most of which is wood (BMU, 2012).

Thus, there is a great potential and also a great need for woody biomass production. The conventional output of forests cannot provide sufficient wood production amounts because of regulatory restrictions and, to an even greater extent, because sustainable usage of forestland is not realistic (Aust, 2012). As a result, cultivation methods such as short-rotation coppice (SRC) have become increasingly important in closing this gap. With this method, willows or poplars are planted on agricultural sites like 'normal' annual cultures. But instead of being harvested after one season, they are allowed to grow for a period of 3–5 years. In total, such plantings remain in use for almost 20 years. Furthermore, different studies have shown that, on the national scale for Germany, SRC has a high potential for being a renewable and decentralized energy carrier (Aust, 2012; Hartwich, Bölscher, & Schulte, 2014a; Murach et al., 2009). Such potential ought to be considered on the European scale.

However, this type of cultivation also has its drawbacks. SRC often requires the reassignment of land and water resources away from their current usage. For example, farmland used for food production may be reassigned to biomass creation for bioenergy. Moreover, SRC has positive and negative effects on land and water issues, especially water quality and quantity. In terms of renewable energy sources, considering these effects in the current process of social reconstruction is essential for reaching the goal of overall sustainable development. This creates a dilemma for the sustainable use of water in the field of renewable energy creation. On the other hand, win–win situations could arise within a proper environmental management concept.

In order to optimize yield and environmental conditions, decision-makers such as land-use planners and private stakeholders (i.e. landowners and farmers) must be provided with information regarding the positive and negative aspects of SRC that affect environmental conditions. This need is further amplified by the variety of perspectives generated by this optimization process in decision-making.

This process gives rise to the following questions. Which positive or negative effects are important and need to be taken into account in the land-use planning and later the cultivation process? Moreover, which challenges arise and have to be considered in the establishment process? These questions relate directly to the aim of this approach: to provide decision-makers and stakeholders with the relevant information about the land and water aspects of SCR. To fulfil this aim, this approach offers a framework of conditions to prevent a dilemma in sustainable water use and to initiate a win–win situation through the establishment of SRC.

To answer these central questions and achieve this aim, this study first introduces the hydrological aspects of SRC cultivation. Next, the focus of this approach shifts to the main aspects of environmental management and considers the effectiveness of SRC in terms of land and water management. As a general rule, the usage of water and land in cultivation practises must be efficient to guarantee sustainability. To address this, the efficiency of SRC will be studied in this approach. Moreover, to realize successful cultivation in relation to water management and ecologic benefit, this study suggests a way to gain establishment and identifies the profiteers of such an establishment. By way of literature research, this approach reflects different environmental management aspects of SRC cultivation and their relevancy to decision-makers and stakeholders.

Conceptually, this study is oriented towards this special issue, in which the editors are seeking to establish a new vision of environmental frontiers. This framework allows one to concentrate the discussion on the physical boundaries between land and water. Consequently, this work considers the vertical frontier and focuses on the influences of SRC establishment on the water cycle (Hartmann & Spit, 2014).

Hydrological aspects of SRC cultivation

This section provides information about the impacts of SRC on the water balance, especially concerning groundwater recharge. The recharge of groundwater is, in general, controlled by aspects of the landscape, such as climate, geology, topography, soil, vegetation and land use (Healy, 2010). The last two-mentioned characteristics, vegetation and land use, influence evapotranspiration and interception, which are the dominant aspects that affect groundwater recharge. Evapotranspiration is defined as the total of components that evaporate from plants and soil, and transpiration is defined as the water used by plants to develop biomass. Interception describes the water content that is stored on the outer vegetation layer (Dimitriou, Busch, Jacobs, Schmidt-Walter, & Lamersdorf, 2009a). All these landscape characteristics are strongly influenced by a change of cultivation from annual crops to SRC (Petzold, Feger, & Schwärzel, 2009; Webb et al.,

2009). Such a dramatic change may result in a decrease of percolation to the aquifer and could trigger a lowering of the groundwater level (BUND, 2010).

Due to canopy development and a large leaf surface area, plants like willow or poplar are able to intercept much more precipitation than shorter types of foliage plants during the vegetation period (Dimitriou et al., 2009a; Nisbet, 2005). Similar to other woody plants, and in comparison with annual crops, there is a higher loss to interception, which also persists at a lower amount during the dormant period in winter. Values of intercepted precipitation during the vegetation period have been reported: Ettala (1988) recorded 31% interception for willow SRC in Finland; Hall (1997) reported 21% interception for poplar in the UK. Due to variations in stand ages, canopy heights or diverse local conditions, these values tended to cover a large range. But in contrast to conventional field crops, with an interception of about 15%, the reported amounts for SRC are significantly higher (Dimitriou et al., 2009a; Hall, 2003). These higher levels of interception lead to a decrease in the water amounts that are reaching the ground, or, in other terms, it lowers the amount of effective rainfall.

Poplar and willow are efficient plants in terms of biomass production, which is linked to a high water use. This results from their natural environment on floodplains or wetland areas, where water availability is high. Moreover, a rapid growth dynamic is essential in these habitats, where physical events like flooding or sediment transport present much stronger limitations to growth. However, their water-use increases even more in warm climates (Dimitriou et al., 2009a). Dimitriou et al. (2009a) and Busch (2009) compared the results of nine different studies that estimated the evapotranspiration amounts for poplar and willow SRC in Sweden, Great Britain and Germany. On average, the test sites reached precipitation amounts of 700 mm during one year. Two-thirds of this amount were lost to evapotranspiration. In comparison with Penman open water evaporation, SRC shows higher values. A similar effect is reported for a crop coefficient, which reflects the evapotranspiration in comparison with a grass-covered site. When compared with grassland, SRC areas reached 20% higher rates of evapotranspiration (Allen, Pereira, Raes, & Smith, 1998; Dimitriou et al., 2009a; Hartwich et al., 2014a; Nisbet, Thomas, & Shah, 2011).

A high water abstraction from the ground, due to a well-developed root system, results in soil with lower moisture content compared with regular field sites (Lamersdorf et al., 2010). This requires a rain event with a longer time span and a higher water content to fill up the porous space in the soil. Following this saturation phase, a percolation to the aquifer takes place, which recharges the groundwater. On the other hand, tillage needs to be done only once before the establishment of an SRC. Consequently, a consistent pore system exists, which supports the percolation process. The above-mentioned aspects make it difficult to determine the influences on leaching when compared with other arable crops (Dimitriou et al., 2009a). However, poplar and willow are able to reach groundwater with roots growing to depths up to 2 m (Volk, Abrahamson, & White, 2001). If the groundwater table is accessible to the plants, they could expand their water use. Nisbet et al. (2011) report a doubling of water use compared with locations without access to groundwater. Furthermore, it is mentioned that the volume of groundwater recharge can be reduced by 50% in comparison with sites where grassland is established.

Nevertheless, due to their impact on the water balance, and because of the general management practices of SRC, the cultivation of these products has a positive impact on water quality (EEA, 2008). With regard to ex-agricultural land with a high nutrient content resulting from previous fertilization, the establishment of SRC is able to lower the fertilization leakage to ground or surface water bodies (Dimitriou et al., 2009a; Nisbet,

2011). Such an effect is also possible in the case of heavy metals, but Dimitriou et al. (2009a) have acknowledged a lack of research in that field. However, in comparison with other arable crops, the need for physical and chemical treatment is reduced to a minimum in SRC areas. Tillage and pest and weed control are only applied during the plant establishment, which results in an extremely low impact on water quality. This becomes even clearer when it is taken into account that the lifespan of this crop varies between 10 and 20 years (Dimitriou et al., 2009a).

Different perspectives regarding the establishment of SRC

To provide decision-makers and stakeholders with relevant information about SRC and its influences on the landscape it is necessary to focus on the different aspects of environmental management. These different aspects rise from the variety of spheres existing in the environmental management decision-making process. Furthermore, the consequences of cultivation, an aspect of land and water management, are also influencing these spheres. These influences and their consequences are studied in this approach by means of a literature review.

The literature reviewed and presented herein was identified as suitable for describing the influences of SRC and their relevance in the decision-making process. This paper is structured around perspectives of decision-making, derived from Hartmann and Needham (2012) and Hartmann and Spit (2014), which emphasize inter alia the role of effectiveness and efficiency in decisions on scarce natural resources. This structure highlights the main subjects of this paper, which relate to the reassignment of land and water by SRC and the effectiveness of that process. Furthermore, the efficiency of that cultivation is presented, reflecting the situation from a land- and water-use perspective. The mode of establishment is considered, as it is a key factor in implementing SRC and increasing its positive benefits while limiting its negative consequences on water management and the ecology. In addition, an important element in the decision-making process is the framework in which the profiteers of that concept are arranged.

These main subjects foster both a scientific and a management perspective and aim at analysing, reflecting upon and evaluating the advantages and disadvantages of SRC and its influence on land and water issues. By providing pertinent information for decision-makers and stakeholders, this study can facilitate the future reassignment of land and water resources for generating renewable energy through biomass. This study also aims to counteract the dilemma between sustainable water use and renewable energy production by identifying win–win situations.

Effectiveness of SRC

Effectiveness is the ratio between the forecasted and actual achievement of objectives. When applied to the reassignment of land and water resources, effectiveness involves a scale for measuring cultivation and water. However, in this study it is necessary to distinguish between primary and secondary objectives. The primary aim of SRC is to establish a renewable energy source triggered by an economic model. Of equal importance, the secondary aim of SRC is to provide ecological benefits. Such results can be achieved by controlling the effects of water abstraction by the plants. Further results may include enhancing bird habitats or minimizing land consumption, thus counteracting the 'food versus fuel dilemma'.

For the primary objective, the first link in the system is the farmer. According to Murach et al. (2009), a farmer would produce willow SRC if the harvestable amounts were over 8 odt ha^{-1} a^{-1} (odt = oven dry tons of biomass). At that amount, the production would be economical and could compete with other agricultural products. Additionally, the primary objective is to find a renewable energy source that reduces CO_2 emissions. Therefore, it becomes obvious that woody biomass is a promising renewable energy carrier. In contrast to a fossil energy source, SRC biomass does not emit any 'new' CO_2 into the atmosphere. However, it should be taken into account that farming and transporting the product produces emissions that cannot be compensated by the biomass production of these plants (Cocco, 2007; Dimitriou et al., 2009a; Heller, Keoleian, Mann, & Volk, 2004; Styles & Jones, 2007).

Unlike the primary objective, the secondary objectives for the establishment of SRC are more closely to the management of land and water resources. Thus, they influence both ecological and social issues. This link becomes even clearer when considering the growing conflict between biofuel and food production, a conflict in which SRC is taking a special position. In recent years, critics have argued that, in general, biomass production as an energy carrier reduces arable land, lowers harvestable amounts of food and, finally, creates a nutrition deficit, predominantly in developing countries. The last argument relates to higher costs of food due to a reduced supply (Mitchell, 2008; FAO, 2009; Zichy, Dürnberger, Formowitz, & Uhl, 2011). On the other hand, a World Bank report by Baffes and Haniotis (2010) explained that the extreme rise of food prices from 2007 to 2008 mainly resulted from 'speculation' on commodities and, only to a lesser extent, the development of the biofuel market. However, due to its low nutrition needs, SRC does not require special treatments like fertilizer, unlike maize or other crops. For this same reason, it can be planted in fields with a very low agricultural potential. If these low potential areas were primarily used, SRC would avoid conflicts with food production (Gerold, Landgraf, Wolf, & Schildbach, 2009). Furthermore, according to the findings of Biswas (2009), the lack of food production is more a water and food management problem than a problem caused by the availability of land resources.

Regarding the secondary objectives of SRC, or the ecological standpoint, the dominant subjects are water management, land consumption and the variety of habitat structures in a cultivated landscape. In this context, if SRC is not cultivated in monoculture, it is capable of providing the maximum amount of habitat structures by offering the key elements in a biotope network (Baum, Weih, Busch, Kroiher, & Bolte, 2009; Köhn, 2009; Schmidt & Glaser, 2009). If these cultivation structures were established as buffer zones next to aquatic environments, they would also have positive effects on water quality. This would be achieved by both a reduction of water percolation through the soil, which reduces leakage of nutrients, heavy metals or other chemicals, and a minimum of fertilization and chemical applications on the plants (Börjesson & Berndes, 2006; Dimitriou et al., 2009a; Elowson, 1999; Nisbet, 2011). Beyond water quality, the water quantity would be influenced by the high water consumption of the plants. This basic issue of water abstraction leads to diverse effects, especially in cases where the groundwater recharge is affected. Problems occur when the groundwater level decreases and the base flow of creeks declines. In such cases, ecosystems would be harmed, especially during dry summer periods when rivers generate discharge through groundwater components (Dimitriou et al., 2009a; Hall, 2003; Nisbet et al., 2011). To avoid these consequences, it is recommended that SRC should not be applied in landscapes with a negative climatic water balance (Nisbet et al., 2011). On the other hand, lowering the groundwater table can benefit regions with a high level of water saturation. The application of SRC in these areas also provides an approach for lowering general maintenance costs. This refers in particular to technical infrastructures that utilize pumping or drainage practices to adjust the groundwater level according to certain standards. Because of their potential for reducing secondary CO_2 emissions and costs produced by the above-mentioned infra-structures, the secondary or ecological objectives of SRC should be brought into greater focus in further research.

In relation to the primary objectives for establishing a renewable energy source, SRC has the ability to reduce CO_2 emissions and compete with other bioenergy solutions like wind or photovoltaics (Heller et al., 2004). If the secondary objectives are also taken into account, which are mainly related to water issues and other ecological impacts, the complexity of SRC advantages and disadvantages becomes visible, especially in the 'food versus fuel dilemma'. However, the facts suggest that SRC will most likely have a positive impact on the environment. Accounting for both the primary and secondary aims of SRC, this bioenergy model rests on a broad basis in terms of effectiveness.

Efficiency from a land- and water-use perspective

Land and water are rare goods, and the efficiency of their usage depends on the comparison of their 'consumption' against the produced benefits. These produced benefits are not determined exclusively by economic factors, but rather they are tied closely to ecological or more water-related issues. Therefore, the functional capability of an environment becomes an integral part of human well-being (Biswas, 2009; Gleick, 2000). Land and water management ensure that sustainable utilization concepts are implemented to guarantee the present and future use of these resources (Rogers & Hall, 2003). Thus, the efficiency of land and water use is less dependent on economic issues and is more a matter of ecological and sustainable utilization.

The establishment of SRC serves multiple needs, which increases its level of efficiency. With regards to its primary objective, to produce a renewable energy source, SRC is a very sustainable solution, especially in contrast to the use of fossil energy carriers or wood from forests (Cocco, 2007; Heller et al., 2004; Styles & Jones, 2007). This point rests on two principal arguments: (1) The majority of forest timber products have a high quality. Their use for energy production presents an uneconomic scenario in which high-quality raw material is used for energy production without former usage. Moreover, the amount of preferred tree species in conventional forests is too low to serve the needed energy demand (Marland & Schlamadinger, 1997). (2) If a forest is managed properly, it serves a need for high-quality wood. More importantly, a forest is a sustainable and ecological system, which needs time as a central criterion to achieve these objectives. These aspects show that conventional forestry meets both an economic need for high-quality wood products and mitigates long-term environmental issues. Also, considering the terms of a cultivated land-scape, it is essential that landscape functions are being preserved for bioenergy production. This approach is conducive to the establishment of a sustainable land management, particularly in a landscape with clustered environmental functions.

The same applies to the water-use perspective. The relevance of efficiency, or rather sustainability, is connected to multiple issues (Söderbaum & Tortajada, 2011). These water-related issues are defined by Gleick, Christian-Smith, and Cooley (2011) as water quality, productivity, reliability and energy demands. As shown previously, the influences of SRC on water quality are primarily positive compared with the influences of other agricultural crops. Furthermore, research indicates that, when planted as buffer strips along fields in the vicinity of aquatic ecosystems, SRC enhances water quality

(Börjesson & Berndes, 2006; Dimitriou et al., 2009a; Elowson, 1999; Nisbet, 2011). The productivity yield of SRC is often estimated by using the water-use efficiency of the tree species planted. Additionally, an estimation of the plant-available water in the environment must be calculated. Different assessment strategies have been developed for this purpose, but in general a close proximity to groundwater is essential for an optimal yield because of the high water consumption of SRC. If SRCs are able to reach the groundwater level with their root system, this establishes optimal conditions for water availability. SRC productivity also refers to the fact that this type of cultivation does not generally require fertilization operations, due to economic conditions and the physiological structure of the plants. The nutrition needs and limited yield prices of the plants are too low to justify fertilization (Aust, 2012; Hartwich et al., 2014a; Murach et al., 2009). In the case of reliability, if the water amounts are restricted in the landscape because of a negative water balance or other issue, it is possible that there will be adverse effects on the biomass production and on the environment. In this context, reliability is defined by Hashimoto, Stedinger, and Loucks (1982) as the extent to which a system is in a suitable state. When this idea is transferred to an environmental management concept, it implies the need for a sustainable selection of locations where the water requirements will be met. If such regulations are not introduced, the system would lose balance and, with it, sustainability. However, if the required energy demands of SRC are met, it seems that a positive energy balance in the cultivation process will be the result. These findings are confirmed by several studies in the field (Cocco, 2007; Heller et al., 2004; Styles & Jones, 2007). However, these studies also suggest that frameworks of components like transportation, cultivation and harvesting are able to lower the general energy output. In order to achieve optimal efficiency and sustainability in land and water usage, it is necessary that these components of energy efficiency be addressed economically.

A sustainable treatment could be achieved if recommendations or regulations were used to set a framework to minimize negative impacts on the environment and optimize sustainable development in ecological terms. This approach would especially support the development and expansion of ecosystem functions in a cultivated landscape.

A way to gain an establishment for SRC

To enable the establishment of SRC, it is necessary to generate and use a win–win situation to trigger positive development in water and ecological issues as well as socio-economic benefits. Pain et al. (1996) published a solution for avoiding conflicts between water resource management, food and energy crops that focused on cultivation of SRC on (1) land with a highly erosive quality, (2) agricultural land with a high water content and (3) land that is not fertile enough for effective food production. This approach stands with other studies that not only count the direct monetary profit of SRC but also consider the environmental benefits (Duer & Christensen, 2010; Hanegraaf, Biewingam, & van der Bijl, 1998; Uchida & Hayashi, 2012). These studies highlight the benefits of bioenergy by focusing on environmental objectives like the improvement of water quality, the increase of habitat variety and the reduction of greenhouse gas emissions. Since these often water-related benefits are linked to the well-being of society, it is logical that SRC should gain the support of subventions. Currently, direct monetary subventions are not established, however indirect support systems can be found. In this context, it is possible to adapt SRC to so-called 'greening activities'. These activities are part of the 'common agricultural policy' (CAP). As such, they use monetary benefits to encourage farmers to establish agricultural sites with high ecosystem functions or qualities to counteract climate change (Busch, 2012; Schmidt et al., 2014). The system of emission trading generates a further indirect subvention for SRC by improving the energy production from renewable sources (Köhn, 2009).

However, the community has to set up stimuli to trigger the developments of SRC so that it can serve society in the broadest sense. As with all economic concepts, monetary incentives generate a strong vehicle for the dispersion of innovation. This applies very well to the innovation of SRC. But a pure monetary incentive would result in an uncontrolled spread of SRC without regard to the social and environmental risks or benefits of its cultivation. Subventions can lower these risks if they are channelling SRC efforts to useful sites. On the other hand, an approach that is not economically based could also provide a good solution to this problem. Such an idea could be based on the above-mentioned 'greening activities' and emission trading, which are also related to ecological treatment of a landscape. In this case, SRC would receive a 'formal' or an economic weight that indicates an intangible value linked to the positive environmental effects of SRC.

Profiteers of cultivation

One of the central questions of this discussion is who benefits from SRC and to what extent do they profit. To answer this question, the actors and their roles must first be identified. The owner of an SRC takes the position of the primary economic beneficiary who profits directly from the harvest. Due to the owner's position in the establishment of SRC, his profit has a strong connection to the market price of wood chips and the costs of cultivation. In this way, and due to his ability to plant, harvest and earn money from planting, the farmer is the main stakeholder in this sector (Pretzsch & Skodawessely, 2010).

The benefit a society receives is based on the level of sustainability that is created by a well-established SRC. In this case, 'well established' refers to the positive aspects described in the former subsections. If these aspects are taken into account, a maximum sustainability is achieved by way of cultivation, which refers to the most positive characteristics. Such an approach also takes the environment into account or, in a broader context, nature. The environment stays preserved due to a sustainable treatment. This benefit counts even more in a cultivated landscape, which is hardly clustered into ecosystem functions. SRC provides the opportunity to re-establish these functions in relation to the discussed water issues. This wide range of ecological profit is expressed by several discussions and studies (Dimitriou et al., 2009b; Schägner, 2009; Schmidt & Gerold, 2010; Schmidt & Glaser, 2009).

If society and nature can be described as stakeholders in this context, then profits can be attributed to them. Schägner (2009) formulated a list of profits and indirect benefits of SRC cultivation which included: (1) a contribution to climate protection due to carbon storage and the lessening of greenhouse gas emissions, (2) positive regulatory effects on the environmental water balance, (3) flood control by water retention, (4) improvement of microclimate and air quality due to the evapotranspiration characteristics of SRC, (5) enhancement of soil and water quality by limiting erosion and fertilizer application, and (6) improvement of habitat variety and ecosystem functions.

Discussion

The establishment of SRC can be seen as a consequence of the new European energy policy. Due to the high volume of timber needed for a future of renewable electricity rather than heat production, the conventional forestry model cannot provide the needed amounts of wood while also remaining sustainable (Aust, 2012). As an alternative, SRC is a type of cultivation associated with the reassignment of land and water resources away from their current usage. Furthermore, SRC may have positive and negative effects on the environment, which should be considered in the planning and cultivation process, and which may counteract a possible dilemma between sustainable water use and renewable energy. From an environmental management standpoint, the effectiveness and efficiency of SRC mainly affects water-related issues.

With regard to the primary objective to establish a renewable energy source, SRC is able to compete with other bioenergy solutions, like solar power or wind. Furthermore, SRC has the ability to reduce CO_2 emissions, if framework parameters like cultivation and transportation of the product are set up properly (Heller et al., 2004).

As another benefit, in a cultivated landscape, SRC is capable of adopting ecosystem functions and enhancing the habitat variety. Moreover, it is reported that a cultivation has positive effects on water quality, especially when these plantings are established in the vicinity of drainage systems or between rivers and agricultural areas (Börjesson & Berndes, 2006; Dimitriou et al., 2009a; Elowson, 1999; Nisbet, 2011).

As to its impact on the ecosystem, SRC is able to reduce groundwater recharge up to 50%, due to its high water abstraction (Nisbet, 2011). This ability could have negative effects (e.g., for the base flow of creeks during summer months, or in generally dry periods, this behaviour would stress the ecosystem). Due to this effect, a cultivation of SRC should not be applied in areas with a negative climatic water balance (Dimitriou et al., 2009a; Hall, 2003; Nisbet et al., 2011). Alternatively, SRC could be used in agricultural areas where the soil has a very high water content and/or groundwater level. This would avoid the necessity of maintaining cost- and labour-intensive drainage and channel systems, such as those used in the cultivated peat and wetlands in Northern Germany and the Netherlands. Furthermore, the ability of SRC to flourish in high water areas could be used to manage high groundwater levels in the vicinity of urban areas as well as to reduce costs for technical groundwater management. Perry, Miller, and Brooks (2001) described a further positive effect of SRC: the reduction of peak water levels in a medium flood event. The water retention capacity of SRC could be used in decentralized flood protection concepts. Bölscher, Schulte, and Huppmann (2010) have shown that in riparian areas of the Upper Rhine, a retention effect not only occurs in the landscape, but also is generated by young, flexible willows on the floodplain.

If SRC was cultivated in the above-mentioned situations and implemented in environmental management strategies, it could serve effectively in multiple water-related needs. SRC could also improve the efficiency of water use, which is closely associated with sustainability, which in turn is related to the issues of quality, productivity, reliability and energy (Gleick et al., 2011). Taking all these issues and aspects into account, SRC is presented as an efficient cultivation practice, as long as it is implemented in a sustainable management concept.

Conclusions

A sustainable management concept must take the positive and negative effects of SRC into account, thus considering the challenges arises of its establishment. In order to realize such a sustainable cultivation, the potential areas for cultivation should be identified to provide an optimal spatial basis for decision-making in terms of environmental benefit. The productivity of SRC is related to water potentials in the landscape and the water-use efficiency of the plants. This connection necessitates the process of approximating yield amounts in the landscape. If the environment is not suitable, due to insufficient water potentials, the expected yield will be low (Aust, 2012; Hartwich et al., 2014a; Murach et al., 2009). This also shows a link to reliability, which is defined as the extent to which a system acts sustainable (Hashimoto et al., 1982). With this in mind, the question arise of how investigations could provide information about sustainable locations where the water requirements for SRC will be met.

In addition to considering the above-mentioned landscape potentials, it is essential to set up stimuli to trigger the development of SRC. A combination of both aspects would help to reduce an uncontrolled spread of SRC. Such a spread could result from a purely economic model of establishment, which does not consider the risks and rewards for society. A strategy using subventions could lower these risks by focusing the SRC efforts on useful sites. 'Greening activities', as a part of the European 'common agricultural policy', could also facilitate the spread of SRC by encouraging farmers to establish SRC to enhance ecosystem functions or rather the shown water-related issues, as well as to counteract climate change (Busch, 2012; Thomas et al., 2014). This establishment concept would also serve the profiteers, with the farmer acting as the main stakeholder through monetary benefit, followed by society and nature, who benefit on the level of sustainable land and water usage (Pretzsch & Skodawessely, 2010; Schägner, 2009).

The challenges of establishing SRC are most highly related to water issues in sustainable cultivation. These water issues can only be taken into account by establishing win–win situations that serve both the monetary and the environmental aspects. Current research uses Germany to show the spatial distribution of high-potential areas for SRC and locates priority areas on the northern German plain (Aust, 2012; Hartwich et al., 2014a, 2014b; Murach et al., 2009). These water-based yield approaches show the strong linkage between water management and yield optimization, which encourages new perspectives in the field of water energy. In addition, these spatial approaches offer information to counteract the possible dilemma of sustainable water use and energy production.

Funding

This work was supported by the Federal Ministry of Food and Agriculture, Germany and Fachagentur Nachwachsende Rohstoffe e. V. (FNR) [grant number 2012410].

References

Allen, G., Pereira, S., Raes, D., & Smith, M. (1998). Crop evapotranspiration – Guidelines for computing crop water requirements – FAO Irrigation and drainage paper 56. Rome: FAO. Retrieved from http://www.fao.org/docrep/x0490e/x0490e00.htm

Aust, C. (2012). Abschätzung der Nationalen und regionalen Biomassepotentiale von Kurzumtriebsplantagen auf landwirtschaftlichen Flächen in Deutschland (Assessment of the national and regional biomass potential of short rotation coppice on agricultural land in Germany). Dissertation. Fakultät für Forst- und Umweltwissenschaften. Albrecht-Ludwigs-Universität. Freiburg im Breisgau, Germany.

Baffes, J., & Haniotis, T. (2010). Placing the 2006/08 Commodity Price Boom into Perspective. Policy Research Working Paper 5371. Washington, DC: World Bank. Retrieved from http://www-wds.worldbank.org/servlet/WDSContentServer/WDSP/IB/2010/07/21/000158349_20100721110120/Rendered/PDF/WPS5371.pdf

Baum, S., Weih, M., Busch, G., Kroiher, F., & Bolte, A. (2009). The impact of short rotation coppice plantations on phytodiversity. *Landbauforschung – vTI Agriculture and Forestry Research, 3*(59), 163–170.

Biswas, A. (2009). Water management: Some personal reflections. *Water International, 34*(4), 402–408. doi:10.1080/02508060903396189

Bölscher, J., Schulte, A., & Huppmann, O. (2010). Long-term flow field monitoring at the Upper Rhine floodplains. In A. Dittrich,K. Koll,J. Aberle, & P. Geisenhainer (Eds.), *River Flow.*

Volume 1. Proceedings of the International Conference on Fluvial Hydraulics (pp. 477–485). Braunschweig: Bundesanstalt für Wasserbau.

Börjesson, P., & Berndes, G. (2006). The prospects for willow plantations for wastewater treatment in Sweden. *Biomass and Bioenergy, 30*, 428–438. doi:10.1016/j.biombioe.2005.11.018

Bundesministerium für Ernährung, Landwirtschaft und Verbraucherschutz [BMELV] & Bundesministerium für Umwelt, Naturschutz und Reaktorsicherheit [BMU]. (2010). Nationaler Biomasseaktionsplan für Deutschland. Beitrag der Biomasse für eine nachhaltige Energieversorgung (National biomass plan Germany. Contributions of biomass in order of a sustainable energy supply). Berlin: BMELV & BMU. Retrieved from http://www.bmel.de/SharedDocs/Downloads/Broschueren/BiomasseaktionsplanNational.pdf?__blob=publicationFile

Bundesministerium für Umwelt, Naturschutz und Reaktorsicherheit [BMU]. (2012). Erneuerbare Energien 2011. Daten des Bundesministeriums für Umwelt, Naturschutz und Reaktorsicherheit zur Entwicklung der erneuerbaren Energien in Deutschland im Jahr 2011 auf der Grundlage der Angaben der Arbeitsgruppe Erneuerbare Energie-Statistik (AGEE-Stat) (Renewable energy 2011. Data of the Federal Ministry for the Environment, Nature Conservation, Building and Nuclear Safety to develop renewable energy in Germany in the year 2011 based on information of the work group renewable energy-statistics). Stuttgart: Zentrum für Sonnenenergie- Wasserstoff-Forschung Baden-Württemberg. Retrieved from http://www.hk24.de/linkableblob/hhihk24/innovation/downloads/1897454/.3./data/Entwicklung_der_Erneuerbaren_Energien_in_2011_Stand_Maerz_2012-data.pdf

Bund für Umwelt und Naturschutz Deutschland [BUND]. (2010). Kurzumtriebsplantagen für die Energieholzgewinnung – Chancen und Risiken (Short rotation coppice to generate fuelwood advantages and disadvantages). Retrieved from http://www.bund.net/fileadmin/bundnet/publikationen/landwirtschaft/20100714_landwirtschaft_bund_position_55_KUP.pdf

Busch, G. (2009). The impact of short rotation coppice cultivation on groundwater recharge – a spatial (planning) perspective. *Landbauforschung – vTI Agriculture and Forestry Research, 3* (59), 207–221.

Busch, G. (2012). Agrarholz auf Ackerflächen – eine multikriterielle Bewertung des Einflusses auf Landschaftsfunktionen. Klimawandel: Was tun!. IALE-D-Jahrestagung 2012, 24.–26. Oktober 2012, Eberswalde, INKA BB – Klimawandel und Innovation, 148–150. Retrieved from http://www.iale.de/fileadmin/user_upload/PDFs/IALE-Jahrestagung_2012/IALE_Tagungsband_2012_Eberswalde.pdf

European Environmental Agency [EEA]. (2008). A review of the possible impact of biomass production from agriculture on water: background paper for the conference 'WFD meets CAP – Looking for a consistent approach'. Retrieved from http://icm.eionet.europa.eu/ETC_Reports/Biomass_WFD_report_V7_final260108-2.pdf

Food and Agricultural Organization of the United Nations [FAO]. (2009). The State of Food Insecurity in the World. Economic crises – impacts and lessons learned. Rome: FAO. Retrieved from: http://www.fao.org/docrep/012/i0876e/i0876e00.htm

Cocco, D. (2007). Comparative study on energy sustainability of biofuel production chains. *Proceedings of the Institution of Mechanical Engineers, Part A: Journal of Power and Energy, 221*, 637–645.

Dimitriou, I., Baum, C., Baum, S., Busch, G., Schulz, U., Köhn, J., ... Bolte, A. (2009b). The impact of short rotation coppice (SRC) cultivation on the environment. *Landbauforschung – vTI Agriculture and Forestry Research, 3*(59), 159–206.

Dimitriou, I., Busch, G., Jacobs, S., Schmidt-Walter, P., & Lamersdorf, N. (2009a). A review of the impacts of short rotation coppice cultivation on water issues. *Landbauforschung – vTI Agriculture and Forestry Research, 3*(59), 197–162.

Duer, H., & Christensen, P. (2010). Socio-economic aspects of different biofuel development pathways. *Biomass and Bioenergy, 34*, 237–243. doi:10.1016/j.biombioe.2009.07.010

Elowson, S. (1999). Willow as a vegetation filter for cleaning of polluted drainage water from agricultural land. *Biomass and Bioenergy, 16*, 281–290. doi:10.1016/S0961-9534(98)00087-7

Ettala, M. O. (1988). Evapotranspiration from Salix aquatic plantation at a sanitary landfill. *Aqua Fennica, 18*, 3–14.

Gerold, D., Landgraf, D., Wolf, H., & Schildbach, M. (2009). Bewirtschaftungsstrategien von Kurzumtriebsplantagen (Strategies of cultivation in terms of short rotation coppice). In T. Reeg,A. Bemmann,W. Konold,D. Murach, & H. Spiecker (Eds.), *Anbau und Nutzung von Bäumen auf landwirtschaftlichen Flächen* (pp. 73–82). Weinheim: Wiley-VCH.

Gleick, P. (2000). The changing water paradigm – a look at twenty-first century water resources development. *Water International, 25*(1), 127–138. doi:10.1080/02508060008686804

Gleick, P., Christian-Smith, J., & Cooley, H. (2011). Water-use efficiency and productivity: Rethinking the basin approach. *Water International, 36*(7), 784–798. doi:10.1080/02508060.2011.631873

Hall, R. L. (1997). Water use of poplar clones grown as short-rotation coppice at two sites in the United Kingdom. *Aspects of Applied Biology, 49*, 163–172.

Hall, R. L. (2003). Short rotation coppice for energy production – Hydrological guidelines. Report of the DTI New and Renewable Energy Programme, Centre for Ecology and Hydrology. Retrieved from: http://nora.nerc.ac.uk/2631/1/HallShortRotationReport.pdf

Hanegraaf, M., Biewingam, E., & van der Bijl, G. (1998). Assessing the ecological and economic sustainability of energy crops. *Biomass and Bioenergy, 15*(4–5), 345–355. doi:10.1016/S0961-9534(98)00042-7

Hartmann, T., & Needham, B. (2012). Introduction: Why reconsider planning by law and property rights? In T. Hartmann & B. Needham (Eds.), *Planning by law and property rights reconsidered* (pp. 1–23). Farnham, Surrey: Ashgate.

Hartmann, T., & Spit, T. (2014). Editorial: Frontiers of land and water governance in urban regions – questions of sectors, scale, and shifts of working paradigms. Water International – this issue.

Hartwich, J., Bölscher, J., & Schulte, A. (2014a). Die Menge des pflanzenverfügbaren Wassers als Kriterium zur Abschätzung des Bioenergiepotentials von KUP-Standorten in Deutschland (The amount of plant available water as criteria to predict biomass potentials of SRC in Germany). In P. Chifflard, D. Karthe, S. Grashey-Jansen, K-F. Wetzel(Eds.), Beiträge zum 45. Jahrestreffen des Arbeitskreises Hydrologie vom 21.–23. November 2013 in Augsburg (pp. 29–34). Geographica Augustana 16.

Hartwich, J., Bölscher, J., & Schulte, A. (2014b). Wasserverfügbarkeit als Kriterium zur Potenzialabschätzung für die Produktion holzartiger Biomasse in Deutschland (Available water as criteria to estimate potentials of woody biomass production in Germany). Poster session Tag der Hydrologie. Eichstätt. Retrieved from: http://www.geo.fu-berlin.de/geog/fachrichtungen/angeog/projekte/AGENT/Hartwich_Boelscher_Schulte_Poster-TdH-2014.pdf?1405009162#%20Hartwich_Boelscher_Schulte_Poster-TdH-2014

Hashimoto, T., Stedinger, J., & Loucks, P. (1982). Reliability, resiliency, and vulnerability criteria for water resource system performance evaluation. *Water Resources Research, 18*(1), 14–20. doi:10.1029/WR018i001p00014

Healy, R. W. (2010). *Estimating groundwater recharge*. Cambridge, UK: The Cambridge University Press.

Heller, M., Keoleian, G., Mann, M., & Volk, T. (2004). Life cycle energy and environmental benefits of generating electricity from willow biomass. *Renewable Energy, 29*, 1023–1042. doi:10.1016/j.renene.2003.11.018

Köhn, J. (2009). Socio-economics in SRC – a review on concepts and the need for transdisciplinary research. *Landbauforschung – vTI Agriculture and Forestry Research, 3*(59), 223–232.

Lamersdorf, N., Petzold, R., Schwärzel, K., Feger, K.-H., Köstner, B., Moderow, U., ... Kunst, C. (2010). Bodenökologische Aspekte von Kurzumtriebsplantagen (Aspects of soil ecology regarding short rotation coppice). In A. Bemmann, & C. Knust, (Eds.), *Agrowood. Kurzumtriebsplantagen in Deutschland und europäische Perspektiven* (pp. 170–188). Berlin: Weißensee Verlag.

Marland, G., & Schlamadinger, B. (1997). Forests for carbon Sequestration or fossil fuel substitution? A sensitivity analysis. *Biomass and Bioenergy, 13*(6), 389–397. doi:10.1016/S0961-9534(97)00027-5

Mitchell, D. (2008). A Note on Rising Food Prices. Policy Research Working Paper 4682. Washington, DC: World Bank. Retrieved from: https://openknowledge.worldbank.org/handle/10986/68200

Murach, D., Hartmann, H., Murn, Y., Schultze, M., Wael, A., & Röhle, H. (2009). Standortsbasierte Leistungsabschätzung in Agrarholzbeständen in Brandenburg und Sachsen (Approximation of yield amounts regarding agroforestry in Brandenburg and Saxony). In T. Reeg,A. Bemmann,W. Konold,D. Murach, & H. Spiecker (Eds.), *Anbau und Nutzung von Bäumen auf landwirtschaftlichen Flächen* (pp. 29–40). Weinheim: Wiley-VCH.

Nisbet, T. (2005). Water Use by Trees. Forestry Commission Information Note 65. Edinburgh, UK: Forestry Commission

Nisbet, T., Thomas, H., & Shah, N. (2011). Short Rotation Forestry and Water. In H. McKay. (Ed.), *Short Rotation Forestry: Review of growth and environmental impacts* (pp. 3–36). Surrey, UK: Forest Research Monogaph, 2.

Pain, L., Peterson, T., Undersander, D., Rineer, K., Bartelt, G., Temple, S., … Klemme, R. (1996). Some ecological and socio-economic considerations for biomass energy crop production. *Biomass and Bioenergy, 10*(4), 231–242. doi:10.1016/0961-9534(95)00072-0

Perry, C. H., Miller, R. C., & Brooks, K. N. (2001). Impacts of short-rotation hybrid poplar plantations on regional water yield. *Forest Ecology and Management, 143*, 143–151. doi:10.1016/S0378-1127(00)00513-2

Petzold, R., Feger, K.-H., & Schwärzel, K. (2009). Wasserhaushalt von Kurzumtriebsplantagen (Water balance of short rotation coppice). In T. Reeg, A. Bemmann, W. Konold, D. Murach, & H. Spiecker. (Eds.), *Anbau und Nutzung von Bäumen auf landwirtschaftlichen Flächen* (pp. 181–192). Weinheim: Wiley-VCH.

Pretzsch, J., & Skodawessely, C. (2010). Sozio-ökonomische und ethische Aspekte der Kurzumtriebswirtschaft (Socio-economic and ethical aspects of short rotation coppice). In A. Bemmann,. & C. Knust,. (Eds.), *Agrowood. Kurzumtriebsplantagen in Deutschland und europäische Perspektiven* (pp. 230–242). Berlin: Weißensee Verlag.

Roger, P., & Hall, A. (2003). Effective Water Governance. Global Water Partnership Technical Committee. TEC Background Papers No. 7. Sweden: Elanders Novum.

Schäger, J. (2009). Monetäre Bewertung ökologischer Leistungen des Agrarholzanbausn (Monetary assessment of ecological benefits regarding short rotation coppice). In T. Reeg, A. Bemmann, W. Konold, D. Murach & H. Spiecker (Eds.), *Anbau und Nutzung von Bäumen auf landwirtschaftlichen Flächen* (pp. 171–180). Weinheim: Wiley-VCH.

Schmidt, P., & Gerold, D. (2010). Nachhaltig bewirtschaftete Wälder versus Kurzumtriebsplantagen versus Agroforstsysteme (Sustainable development of forestry vs. short rotation coppice vs. agroforestry systems). In A. Bemmann, & C. Knust (Eds.), *Agrowood. Kurzumtriebsplantagen in Deutschland und europäische Perspektiven* (pp. 208–216). Berlin: Weißensee Verlag.

Schmidt, P., & Glaser, T. (2009). Kurzumtriebsplantagen aus Sicht des Naturschutzes (Short rotation coppice form a environmental conservation perspective). In T. Reeg, A. Bemmann, W. Konold, D. Murach, & H. Spiecker (Eds.), *Anbau und Nutzung von Bäumen auf landwirtschaftlichen Flächen* (pp. 161–170). Weinheim: Wiley-VCH.

Schmidt, T., Röder, N., Dauer, J., Klimek, S., Laggner, A., de Witte, T., … Osterburg, B. (2014). Biodiversitätsrelevante Regelungen zur nationalen Umsetzung des Greenings der Gemeinsamen Agrarpolitik der EU nach 2013 (Regulations of biodiversity in the national strategy of „greening-activities' related to the common agricultural policy after 2013). Braunschweig, Germany: Thünen Working Paper, No. 20. Retrieved from http://nbn-resolving.de/urn:nbn:de:gbv:253-201403-dn053406-9

Söderbaum, P., & Tortajada, C. (2011). Perspectives for water management within the context of sustainable development. *Water International, 36*(7), 812–827. doi:10.1080/02508060.2011.628574

Styles, D., & Jones, M. (2007). Energy crops in Ireland: Quantifying the potential life-cycle greenhouse gas reductions of energy-crop electricity. *Biomass and Bioenergy, 31* (11–12), 759–772. doi:10.1016/j.biombioe.2007.05.003

Uchida, S., & Hayashi, K. (2012). Comparative life cycle assessment of improved and conventional cultivation practices for energy crops in Japan. *Biomass and Bioenergy, 36*, 302–315. doi:10.1016/j.biombioe.2011.10.043

Volk, T., Abrahamson, L., & White, E. (2001). *Root dynamics in willow biomass crops*. New York, NY: SUNY College of Environmental Science and Forestry. United States Department of Energy Syracuse.

Webb, J., Cook, P., Skiba, U., Levy, P., Sajwaj, T., Parker, C., & Mouat, A. (2009). Investigation of the Economics and Potential Environmental Impacts of the Production of Short Rotation Coppicing on Poorer Quality Land. Report to the Scottish Government Retrieved from http://www.scotland.gov.uk/Publications/2009/10/21091129/0

Zichy, M., Dürnberger, C., Formowitz, B., & Uhl, A. (2011). *Energie aus Biomasse ein ethisches Diskussionsmodell. Eine Studie des Institutes Technik-Theologie-Naturwissenschaften und des Technologie- und Förderzentrums (Energy from biomass a ethical discussion model)*. Wiesbaden, Germany: Vieweg + Teubner Verlag, Springer Fachmedien.

Regional governance vis-a-vis water supply and wastewater disposal: research and applied science in two disconnected fields

Martin Schmidt

Spatial and Infrastructure Planning, Technische Universität Darmstadt, Darmstadt, Germany

Water supply and wastewater disposal constitute a key stake in the sustainable development of urban regions. The provision of urban water management is often the responsibility of the municipalities. Currently, extensive governance challenges for the optimizing of water supply and wastewater disposal emerge. Based on an analysis of three German urban regions, the paper argues that there is an increasing need to enter into regional collaboration for the strategic further development of urban water management. From a spatial research perspective, therefore, the so far severely neglected intersection between infrastructure governance and regional governance is elucidated in its various dimensions.

Introduction

The interdependencies of water and space are structured significantly by urban water supply and wastewater disposal. Urban water management provides basic services that are crucial for people's livelihood and at the same time constitute a key interface between nature and society. Particularly in European countries, governance issues are less related to (initial) access to such services, but concern redevelopment, modernization and the further development of water infrastructures. The latter may already have been developed to high standards in order to facilitate the improvement of their efficiency and secure their sustainability in the face of multiple changes. This requires a high level of coordination and close collaboration by the actors involved. This is even more the case for urban regions, which are usually characterized by longstanding traditions and a multiple entanglement of regional collaboration in different areas, in their respective individual spatial layouts and on different levels. In the scientific governance debate the subarea of 'regional governance' addresses this subject and predominantly includes issues of multilevel governance.

With regard to urban water management, various studies have attributed significant relevance to municipalities on the lowest governance level for water supply and wastewater disposal issues. 'At the local level, a mixture of governance and management responsibilities comes into play' and here water governance is being significantly monitored by local political processes or rather the local constituencies (Grigg, 2011, p. 808). As Furlong asserts in the example of a case study of water governance in the Province of Ontario (Canada), the legal and economic provisions in place do not produce the desired impact on higher government levels as long as challenges at the municipal level are rarely examined systematically, neglected or even ignored. Thus good water governance is hardly achievable without good urban governance (Furlong, 2012). In their reflection on the (globally relevant) water governance debate, Mehta et al. (2007) assert that 'critical social science perspectives have emphasized multiple levels (global, local and in-between) [...] and see institutions as part of a constant process of negotiation that involves power and conflicting interests within communities' (p. 27). As a conclusion, they highlight that 'scale remains an issue, with multilevel, networked governance arrangements being an important complement to both global-level and local approaches' (p. 32). It is to be assumed that municipalities are only able to fulfil their critical role in the interplay of spatial and water system development in regional space. Particularly where national law defines municipalities as the main providers, inter-municipal collaboration needs to be systematically established, developed and adjusted to new challenges at the regional level. Single municipalities are hardly able to cope with the increasing constraints on its opportunities for action and control. Collaborating with other municipalities, however, enables them to meet the demands of a sustainable water supply and wastewater disposal.

Research on infrastructure has not yet been adequately linked to research on regional governance. Important relationships and mutual impulses are thus ignored that could be profitable for the sustainable governance of land and water in urban regions. This paper aims at illuminating the interface between regional governance and the governance of urban water management both theoretically and empirically. With regard to the latter, collaboration in urban water management in three German urban regions is analysed and these questions are discussed: What forms and deficits of inter-municipal and regional collaboration can be identified in urban water and wastewater management? What is the connection between this functional collaboration and approaches to regional governance? Based on the empirical evidence on the performance of collaboration so far and its connection to the regional governance of the individual urban regions, the following question is posed: Which conditions shape the development of inter-municipal and regional collaboration and what is the role of spatial planning in this context? This paper examines the importance of water infrastructure for the development of urban regions and identifies the problems involved in the present management forms. At the same time central insights are gained into the major frontiers of regional water governance, and innovation requirements with regard to planning are highlighted.

The paper is structured as follows. First, it reviews the state of the art with a conceptual overview of infrastructure and regional governance research. Second, it analyses the particularities of the water supply and wastewater disposal sector in Germany. Based on a secondary empirical analysis, the socio-technical, contextual and economic changes that present the challenge to develop new concepts of land and water governance are elaborated. Third, concrete empirical findings regarding current urban water management collaboration in three German urban regions are presented. The contribution closes with conclusions for infrastructural and regional governance.

Debates on infrastructure and regional governance

In previous research there is a longstanding debate on the functions, characteristics and interdependencies of technical infrastructure systems. Also, a valuable body of urban and regional research is available, including a debate on the governance of metropolitan regions – based on studies of regionalization.

Characteristics and functions of technical infrastructure systems

Since the second half of the 1980s, the technical and institutional artefacts of large technical infrastructure systems have been subject to explicit consideration in the debate on 'large technical systems' (LTS). The organizational structure of infrastructures has been coined by the technologies implemented, while these systems are at the same time also socially constructed and formed by the institutional context (Monstadt, 2004, p. 33). There are close interdependencies here between material and social components and it becomes clear that 'technologies are shaped by society at the same time as they shape society or, in other words, that (social) technical systems and (technical) societies co-evolve' (Coutard, Hanley, & Zimmerman, 2005, p. 1). Against this background the debate describes in detail the existing path dependencies of large technical infrastructure systems which – in conclusion – are characterized by limitations to attempts at political control (Monstadt, 2007, p. 10).

Another line of infrastructure research entangles with spatial development. Infrastructure policy can normally be seen as economic and structural policy. A (contested) new research approach to this is the thesis of splintering urbanism. With this thesis, Graham and Marvin (2008) assume that infrastructure networks have lost their function of economic and social cohesion and have evolved into a medium of unequal spatial development. Apparently uncontested is the existence of interplay between the transformation of technical networks and the local context with its social characteristics – technological transitions shape space, and spatial conditions influence the development of technology (Hodson & Marvin, 2007, p. 247; Hodson & Marvin, 2009).

A third line of infrastructure research is the analysis of the socio-ecological dimensions of supply and disposal systems. As Kaika and Swyngedouw (2000, p. 120) argue, 'technological networks are the material mediators between nature and the city, [...] the mediators through which the perpetual process of transformation of nature into city takes place'. However, infrastructures as the backbone of sustainable urban development and mediators of urban metabolism have hardly been considered so far (McFarlane & Rutherford, 2008). Thus, it is necessary to review systematically the 'interwoven knots of social process, material metabolism, and spatial form that go into the formation of contemporary urban socionatural landscapes' (Heyen, Kaika, & Swyngedouw, 2006, p. 8).

Regulation and collaboration in urban regions

When examining the discourse on regional governance, it is necessary to begin with the increasing significance of the regional level. This requires discussions of suitable models and organizational structures for urban regions, which are connected to governance aspects of planning activities in institutional, socio-cultural and spatial-specific contexts, and the issues of 'power', 'agency', 'institutions' and 'regulation' need to be clarified. The term 'regional governance' emphasizes the regional level. Here, the discourse combines a theoretically based analysis of regulation and structural forms for the collaboration of urban regions with both formal and informal elements (Benz, 2001, p. 56). The 'governance' stresses the collaboration of government, economic and civil society actors in urban regions.

Two approaches can be observed in the regional governance debate. First, a rather analytical concept is applied to analyse structures and processes in regions. This concept is primarily descriptive and is hardly suitable for evaluating governance forms as particularly advantageous or problematical, or for placing them in connection with policy outcomes. Second, there is a normative approach that seeks to identify forms of 'good regulation'. In contrast to the analytical scientific concept, this approach is more strongly related to practice and seeks to name requirements for the modern management of urban regions. The main subject-matter in all approaches to regional governance is the community orientation of the actors involved. Since the 1990s, or since 'new regionalism' (Lackowska, 2011, p. 27), governance research has assumed that such a community orientation is most likely to evolve when efforts rest on voluntary cooperation and not on imposed collaboration structures.

For scientific purposes the regional governance concept tends to be employed in an analytical sense. A common critique claims that it is of little use to generate statements regarding the ideal form of regulation of a region or even concrete proposals for improvements. 'It seems we are better at explaining failure than identifying factors that will ensure the success of metropolitan reform initiatives' (Heinelt & Zimmermann, 2011, p. 1176, in reference to Basolo). Admittedly, by now insights into individual development factors and areas of collaboration exist, yet there is still no concept to comprehend systematically the governance structures of urban regions (Pütz, 2004, p. 97; Fürst, 2007, p. 358). At the same time, however, the regional governance debate does not sufficiently consider concrete policy content.

Linking the two research perspectives

This brief overview illustrates that the debates stem from different and heterogeneous research in separate disciplines. The lines reveal some research gaps. Monstadt (2009) summarizes the conceptual state of the infrastructure research as insufficient in 'that none of the theoretical approaches discussed is entirely appropriate for conceptualizing the complex interrelation between cities, urban technologies, and natural environments' (p. 1935).

Even larger gaps become apparent when examining the intersection of research on technical infrastructures and on regional governance. So far the individual discourses have provided crucial points of reference for partial issues (see below), but are insufficiently interlinked to facilitate a well-founded conceptual framework for analyses of the regional governance of land and water. For example, infrastructure research does not permit the deduction of statements regarding the need for regulation in terms of infrastructure provision planning in urban regions. Regional governance research, on the other hand, focuses strongly on collaboration in economic areas while infrastructure as an element that structures space, as well as technical aspects, is entirely blanked out.

The intersection of the research lines has so far been under-researched. This paper focuses on this intersection and thus places itself between the two lines of research. Nevertheless, the following findings from the existing literature are relevant and can be used in the further work. First, the *co-evolution* of infrastructure systems and the space in which they are embedded is determined by the social components of the respective infrastructure agents and their institutional context. This provides an indication for the empirical search for frontiers of land and water governance beyond technical factors. Second, the regional governance discussion highlights the *role of spatial planning* for the support of regional collaboration. Particularly, planning at the regional level is assigned engine, initiative and coordination functions (Danielzyk & Rietzel, 2003, p. 521). According to Knieling (2003, p. 477) planning can substantially support modern regional governance by establishing links among actors in the region and by initiating, organizing and moderating inter-sectoral activities. However, to this end, new instruments beyond classical planning would need to be applied (Kilper, 2007, p. 60).

Challenges and collaboration requirements in urban water management

Attention will now be turned to the required linking of regional governance in water supply and wastewater disposal. The challenges to urban water management in European countries therefore need to be further investigated. This paper focuses on the situation in Germany as a whole before going into the case studies. Which features characterizing German urban water management make it an interesting case in point for such an analysis? While there are discussions regarding the organization of water supply and wastewater disposal in several European countries, the issue is particularly significant in Germany. On the one hand, municipalities rank high as key providers of water supply and wastewater disposal due to the constitutional 'guarantee of self-administration'. On the other hand, the municipal responsibility for public water management results in a particularly small-scale organizational structure with a multitude of enterprises. 'Compared with other countries, the German water sector is very fragmented and small-scaled' (Wackerbauer, 2009, p. 5). Germany is divided into 16 federal states and has roughly 12,300 municipalities. It is estimated that there is a total of about 6400 water supply companies and 6900 wastewater companies (ATT et al., 2011), which are distributed very unevenly across the individual federal states.[1] Considerable differences among the states can also be observed in other data (sources of drinking water, water consumption, financing and levies, water prices etc.), so that comparisons are often made within Germany and used for benchmarking the performance of individual companies. The causes of the differences are usually regarded as lying in the influence of external conditions on structural–technical, commercial and cost indicators (Holländer, Fälsch, Geyler, & Lautenschläger, 2009; Holländer, Lautenschläger, Rüger, & Fälsch, 2013). Irrespective of internal diversities, the decentralized organizational structure in Germany faces stronger debates about efficiency criteria than more centralized structures such as, for example, in England/Wales, the Netherlands, Spain and Scotland, but also in Belgium and Denmark. Nevertheless, explicit research on the professionalization of microenterprises in the context of regionalization (Störmer et al., 2009; Truffer, Störmer, Maurer, & Ruef, 2010) is underdeveloped.

Figure 1 shows that Germany has one of the highest proportions of small water suppliers (fewer than 100,000 residents served) compared with other European countries. This small-scale structure with more than 80 water supply enterprises per 1 million people (for comparison: France is one enterprise; cf. Federal Ministry for Economic Affairs and Labour, 2005, p. 5) invokes multiple criticisms.[2] The discussion on enterprise structure began in the mid-1990s when a report by the World Bank examined the institutional structure of the water sector in Germany. The report placed a negative emphasis on the high municipal fragmentation of the sector and its 'insufficient attention to economic efficiency and costs' (Briscoe, 1995, p. 4). The discussions have continued up to the present day, emphasizing, on the one hand, the internationally leading level of supply, for instance with regard to security, high water quality, the nearly nationwide degree of infrastructure provision and low loss rates. For example, Green and Anton (2012) examine water management in Germany and England using selected parameters (e.g. level of per capita consumption, forms of sustainable urban drainage systems such as rainwater harvesting, green roofs etc.). They find that Germany is 'a front runner in the adoption of sustainable urban water management technologies' and that the 'fragmented

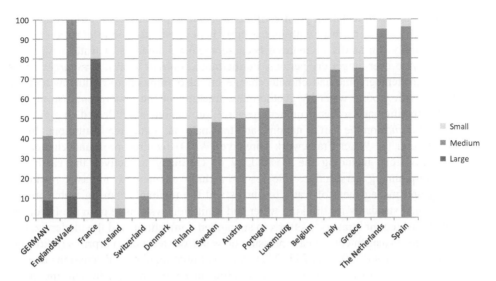

Figure 1. Size structure of water supply enterprises in Europe (2003). Source: Author's own diagram based on Wackerbauer (n.d., p. 4).

system has been much more successful [...] than has the highly centralised system in England' (pp. 208, 209). On the other hand, there continues to be criticism of the high prices in the German system compared with other countries (Wackerbauer, 2007, p. 5). Studies 'found large differences in efficiency, an indication of the potential for cost savings and consumer price decreases. Furthermore, the striking inefficiency of small water utilities introduces the issue of the adequacy of such firms' supply structures' (Zschille & Walter, 2011, p. 20). Current demands often face a lack of qualified staff at the municipal level and the institutional separation of water supply and wastewater disposal.

This also impedes responses to changing socio-economic and ecological restraints that influence the sector and necessitate new modes of governance. Thus, a change in *patterns of demand* can be observed. The population in Germany is undergoing demographic change (an increasing elderly population and a shrinking total size of the population – a reduction of about 17 million people, or 20.7%, until 2060) (Federal Statistical Office, 2009). Particularly in small rural municipalities this is already having considerable consequences for the technical infrastructure. Technical problems are, for example, low consumption, which endangers the quality of drinking water, sedimentation takes place in the channel system, and the functioning of sewage plants is affected. The ensuing countermeasures lead to additional costs. At the same time the operating costs must be met by fewer people, so that infrastructure services become much more expensive. Due to the long life of the plant and the high fixed costs, it is impossible to react in the short-term to the change in the demand situation. The problems are worsened by the fact that water consumption per capita has been decreasing anyway for quite some time (from 147 litres per capita per day in 1990 to 122 litres in 2009). The infrastructure systems, in contrast, were planned at the building stage for a strong increase in water consumption and are often over-dimensioned (German Association of Energy and Water Industries, 2011). The increasing number of the elderly population means an increase in the consumption of pharmaceutical products, which challenge sewerage systems and the environment.

The dismantling of technical plant conflicts with the expectation that larger quantities of sewage will have to be dealt with increasingly heavy rainfalls due to *climate change*. The water supply can hardly be reduced because adequate amounts of water have to be available even in ever-longer dry periods. Furthermore, a *threat to water resources* arises not only from demographic and climate change but also from micro-pollutants and the interference with ground water by geothermics, fracking or the underground storage of gases. The requirements of environment/resource protection are reflected in *tightened regulatory demands* on the part of national and transnational regulative institutions. In the European Union the Water Framework Directive (WFD) required profound changes in statutory provisions including the Basic Law of Germany to deliver the catchment management-planning perspective. The results include an increase in tasks regarding monitoring and management of water resources and the regionalization of water protection. Additional standards for drinking water and sanitation systems (e.g. in the European Union Water Blueprint Strategy[3]), efforts to strengthen competition and the modernization of public sector management increase pressure on water/wastewater enterprises (Allouche, Finger, & Luís-Manso, 2008, pp. 223, 227). Also, national regulatory discussions and the increased activities of cartel authorities cause efficiency pressure. In addition, the enterprises continue to take on new tasks in order to meet higher demands with regard to drinking water quality and its control.

This not only produces administrative demands but also requires *technical innovations*. Thus, in the wastewater treatment an additional purification process needs to be established,[4] the energy efficiency of wastewater treatment facilities optimized and the use of material-energetic potentials of wastewater increased. Also, there is a modernization backlog in urban water management and significant *reinvestment is required* – particularly in the municipal sewer networks and pipelines. Furthermore, the technical and operational development of potential linkages to other supply and disposal sectors needs to be systematically established.

To summarize, German urban water management is highly fragmented, and dominated by municipal enterprises, while a multitude of developments leads to new challenges to facilitate the coordination of water infrastructures in space. The development trends described individually above can be seen as a mélange of technical, legal, social, market, spatial, institutional and resource-related changes. As a result the actors at the local level are faced with new problems and issues of non-technical nature outside their familiar subject area. So, expanded inter-municipal collaboration offers important courses of action and is correspondingly in demand. The German federal government declared collaboration as a key element in modernizing the infrastructure sector (German Bundestag, 2006, p. 4; Wackerbauer, 2009, p. 24).[5] This is in accordance with the intention of the deliberations in this special issue. Primarily technical innovations are not required, rather organizational adjustments including collaboration are needed to take place in all areas from the field of knowledge to activities. The central further development of existing forms of regulation in this connection is their embedding in the respective regional context. Presently, however, in many locations water supply and wastewater disposal – in contrast to social infrastructure or administrative tasks – hardly find themselves at the centre of current collaboration efforts. It is therefore now necessary to place this on an empirical foundation.

Water supply and wastewater disposal in German urban regions

In the following case studies were conducted in three German urban regions (based on a most different case study design[6]): Berlin, including its hinterland, Frankfurt with the

Rhine-Main region, and the Ruhr area. These regions are characterized by strongly accentuated patterns of regional governance and regional water supply and wastewater disposal. They are also particularly significant due to the fact that together they comprise almost one-fifth of the German population. In addition, each of the urban regions has already exhibited close collaborative relationships among the towns and cities in the region since the first half of the 20th century. The qualitative study is based on around 50 expert interviews. The main findings are presented in the following. Since the complete rendering of the analysis of all three regions would take up too much space,[7] core observations will be elucidated with a focus made on the aims of this article – structured according to (1) the region-specific relevance of water infrastructures for the urban regions, (2) an analysis of present collaboration, and (3) the conditions for the development of collaboration.

The co-evolution of land use, regional planning and water governance

Urbanization has always been closely linked to the usability of water. There are therefore also far-reaching indications in the three regions that their growth was coupled to the solution of the water problem. This can be seen most significantly in the urban Ruhr region. A historical review of the Ruhr area shows the co-evolution of land use and water governance: the increasing industrialization and frequent droughts at the beginning of the 20th century led to recurring supply shortfalls and necessitated joint action to secure the water supply. The structures created for this were, on the one hand, a nucleus of collaboration among the individual municipalities. Associations and cooperatives were established that have remained key actors in water management in the region until today. On the other hand, the region and its associations were planned according to the water systems. Water infrastructure systems have subsequently played a major role in shaping the spatial distribution of industrial enterprises and have thus structured spatial development.

At the same time the rivers were extensively changed due to economic and social requirements. Intensive mining resulted in mine subsidence, which meant that it was impossible to implement classic wastewater disposal via underground channels. The River Emscher was therefore used as an open wastewater channel and for decades defined the image of the region as a dirty industrial landscape. As a rule rivers in the region sagged, lay under the natural water level and were furnished with mile-long dykes (Reuther & Schreiber, 2009, p. 193). Today close to 40% of the Emscher region needs to be permanently drained by pumps in order to prevent the flooding of vast parts of the landscape (Emschergenossenschaft & Lippeverband, 2008, p. 6). The extensive polder landscapes have imposed a long-term spatial structure on the urban region and its land use. Even today, when mining has been discontinued, approaches to future regional development are still strongly influenced by the water systems. Meanwhile an underground channel across the whole urban region is under construction (overall 400 km) and the Emscher is undergoing renaturation. Here, the departure from the image of the open sewer constitutes a central determinant in the competition between regions. In this way, water governance issues shape the development of a region with 5 million inhabitants in 53 municipalities. In the process of co-evolution, not only the frontiers of the municipalities need to be bridged but also the frontier between water management and the development of the regional space as habitat and economic zones. Water governance institutions that are more than 100 years old count among the most relevant agents of regional governance.

Performance of collaboration in water supply and wastewater disposal

As shown above, current structures of water resource management are closely tied to decisions in the past and institutions can be traced to previous infrastructure generations. This is also confirmed in the other regions and has a bearing on today's collaboration in water supply and wastewater disposal.

Collaboration structures, contents and deficits

The above finding can be observed most clearly in the urban region of Frankfurt/Rhine-Main. Here the current forms of inter-municipal collaboration can be traced back to the 1960s–70s. At that time secure water supply counted among the most pressing issues of metropolitan development. On the one hand, collaboration for regional water acquisition over large distances developed. On the other hand, specific associations were established for the joint management of wastewater treatment facilities which were mainly aligned with the topography. Altogether, 17 associations are responsible for water acquisition in the urban region with 2.2 million inhabitants in 75 municipalities. Wastewater disposal is split among 85 providers and 19 associations. This spatial and institutional fragmentation results in various deficits which hinder water governance adjusted to efficiency and expected future challenges.

Due to the high number of associations in the water supply and wastewater disposal sector, more formal affiliations of municipalities exist than in other areas of municipal responsibility. The associations, however, have hardly developed structurally since their formation phase even though in the meantime municipal spatial boundaries have been reformed. This means, on the one hand, that the number of members in an associations has decreased to between one and three. On the other hand, different associations are now responsible for individual quarters of the same municipality. Decisions affecting the water infrastructure of a city therefore need to be brought about in different institutions. As a consequence, economies of scale are not put into effect and complex double structures exist. The associations focus on the resolution of (previously) individual issues and individual facilities. The long lifespan and the long funding period of technical facilities cause a path dependency which significantly complicates short-term adjustments.

The fragmentation of the structures is reflected in the fragmentation of the execution of tasks. First of all, due to the structure of the associations water supply and wastewater disposal are organized in separate institutions so that intersections and synergies between the partial branches are not utilized (as the majority of the experts interviewed critically emphasized). In addition, the inter-municipal institutions are not in charge of the entire water supply or wastewater disposal. The bulk of tasks remains with the individual local authorities – e.g. more than half of the wastewater treatment facilities in the region are run by municipalities. Then again, inter-municipal collaboration is only responsible for partial tasks along the value addition chain such as the purification of drinking water or the cleaning of wastewater. This deficit is all the more relevant as particularly the finance and modernization-intensive sub-segments of the extensive network infrastructure (such as water distribution or urban drainage) are under the authority of individual municipalities. Within the entire infrastructure system, the networks of urban water management espe-cially are characterized by high proportions of fixed costs and are particularly vulnerable to future challenges. Even now enormous reinvestment needs are evident in the urban water management networks. The required strategic planning, however, resides with a large number of small towns. But also in existing associations, non-profit agencies or municipal employees are often in charge and these are confronted with both a lack of time and knowledge deficits.

Altogether it seems doubtful whether the existing and strongly inefficient collaboration will be able to compensate the foreseeable problems in urban water management such as the reinvestment needs for sewers etc. mentioned above. The challenges to water management and its interdependencies with spatial development are faced with compartmentalized governance patterns. There is no adequate institutional platform for a higher level coordination of the many and varied individual initiatives, but at best some first approaches. This raises the issue of a regional strategy in the context of regional governance.

Relationship of individual collaboration projects to regional governance

What is the added advantage of a regional governance viewpoint in the context of sectoral collaboration in water supply and wastewater disposal? In order to answer this question it is necessary to resort to the research on regional governance. German exponents of the debate assert that regional governance is able to make sectoral and compartmentalized governance forms more effective if the linkages between them are improved (Fürst, 2003, p. 444). Thus the issue-bound governance forms need to be coordinated into 'regional governance' (Benz & Fürst, 2003, p. 57). It therefore needs to be examined to what extent the mostly locally aligned collaboration is integrated into regional objectives at a higher level and what functions are assigned to it that go beyond the local context.

In the urban region of Frankfurt such a linking can be observed only partially. There is no overall linking of the individual associations on a higher level (Monstadt, Schmidt, & Wilts, 2012). For a structured strategic integration an overall regional regulatory institution and a common objective would be required. In contrast to its predecessor organizations, the existent statutory planning institution of the region (Regionalverband FrankfurtRheinMain, i.e. Regional Association Frankfurt/Rhine-Main) is not responsible for water management. The legislator responsible has deliberately withdrawn these competencies. Issues of water management are regarded as having been satisfactorily solved quite some time ago and must therefore no longer be named as the subject of regional cooperation (State Parliament of Hesse, 2010, p. 18). The governance of the urban region is thus characterized by a focus on economic areas and other thematic discourses beyond the supply and disposal tasks of the municipalities.

In the Berlin region the picture is similar. Here the relationship city–hinterland is hardly perceived as a collaboration space in its own right. Collaboration between the capital and its surrounding municipalities are for the most part tied to single projects while hardly any discussion of urban regional governance takes place nor does a common regional association exist. A rather dated understanding of governance can be found which for the main part manifests itself in the collaboration of state actors in administrative and planning tasks. In the urban region quite intensive water resource management interrelations exist with regard to technical systems and spatial use patterns (as a result of suburbanization effects). Nevertheless, there is hardly any collaboration in water supply and wastewater disposal across the frontier of the metropolis. Initiatives by the Berlin water enterprise to expand business in the hinterland have been declining. There is no overall regional strategy and no concept for a cross-frontier water supply and wastewater disposal. Although water resource management challenges exist, these are not discussed in the overall region but remain left to the devices of the individual enterprises.

The Ruhr urban region, in contrast, demonstrates the advantages of a regional governance strategy. There, the intersection between regional water management and the development of living and economic space has been recognized. The reconstruction of the Emscher and the new alignment of wastewater disposal operate as key factors in regional development. The regional-strategic and higher level integration facilitate not only the interconnection and regionalization that are desirable from a water management point of view (since the WFD aims at the regionalization of water protection as shown above), but are also able to influence positively the further development of regional governance processes.

Frontiers of regional water governance

The assumption has thus been confirmed that collaboration projects that are oriented more strongly to the regional level with a joint strategy are more likely to cope with the challenges to urban water management and the interdependencies between water management and spatial use than are individual, locally limited water supply and wastewater disposal initiatives. It is therefore crucial to analyse the obstacles to increased collaboration. What are the frontiers of regional governance in the water sector? Two observations can be made regarding individual impediments and the role of spatial planning.

In a comparison of the three regions analysed, various *impediments* to inter-municipal collaboration in water supply and wastewater disposal become evident. Technical aspects are relevant here, e.g. when the current condition of the technical facilities differs greatly between municipalities and a potential collaboration partner does not wish to inherit a poorly maintained infrastructure system. Likewise, the minor role played by urban water management issues, the provision of which is taken for granted by the public, becomes apparent. Therefore the local politicians who would be able to take decisions on collaborative activities focus on other issues in order to distinguish themselves for elections. At the same time, the logics of action driven by local egoisms work against regional activities. Collaboration thus essentially depends on the actors and their institutional contexts. Individual persons are always linked to interests in retaining power and mutual personal tensions. In view of the long tradition of the existing institutions in water supply and wastewater disposal, institutional path dependencies have also evolved, however, which are characterized by the competitive situations between municipalities and enterprises. Here, already existing collaboration projects have developed an existence of their own which shows distinct knowledge deficits with regard to professional and strategically directed collaboration.

With regard to the respective regulation, *spatial planning* at the regional level is attributed a significant role as seen in the research state of the art. The planning potential here lies in the interface function between the individual actors in the region, so that planning is able to assume the function of initiating and moderating in order to strengthen regional collaboration. However, urban water management in the regions analysed is often only a remote aspect in regional planning documents. Across the board no concrete statements which would encourage collaboration in this field can be found. All the more important is the cooperation of the regional planning institution beyond formal planning. This seems to be the most likely case in the Ruhr urban region which already displayed the most far-reaching interlinking of water and regional governance. Still, in all regions deficits in the systematic analysis of challenges in urban water management and the corresponding integrated solution strategies are manifest. With a view to the competency to regulate location issues on an overall regional level, spatial planning needs to enter a dialogue process with the supply and disposal enterprises. In the consistent findings it becomes clear, however, that the key frontiers in this area are constituted by the insufficient interlinking of cross-sectoral, coordinating spatial planning with the sectoral planning of water management. If water governance is to be an element of regional governance, urban development needs to be adjusted to the requirements of the technical infrastructure, which presupposes the permanent strengthening of the meta-sectoral coordination competency by spatial planning.

Summary

Beginning with the observation that in the multilevel governance system the municipalities are highly relevant for sustainable urban water management, and that particularly in urban regions the corresponding challenges can be ideally resolved through operational inter-municipal collaboration, various aspects related to the interface of regional and water governance have been examined. Interest was directed both to the theoretical–conceptual framework and to empirical findings regarding collaboration in water supply and wastewater disposal as well as their development factors.

First, it was possible to observe that the relevant research approaches are insufficiently interlinked. The scientific debates on technical infrastructure systems and on the regulation of collaboration between different actors in urban regions have so far not been adequately coupled with the result that main aspects are often not considered together. Drawing from this, it could be seen in the example of challenges to German urban water management – which is generally attributed with a comparatively high standard – how urgent even in European countries organizational issues, and particularly collaboration between the individual municipal providers, are in order to secure an efficient and sustainable water supply and wastewater disposal. Through empirical evaluations of the situation it became apparent how closely the development of urban regions has always interacted with regional water management. Water supply and wastewater disposal have therefore been a major subject of inter-municipal collaboration.

However, many governance challenges are related to the current situation. The existing collaboration projects are in many cases positioned in dysfunctional ways because their institutional structure, spatial layout and the practised task distribution do not match the discernible mesh of problems. Altogether it can be concluded that it is not the establishment of new institutions or new collaborations that is the order of the day – as is often claimed – but that the existing collaboration should be restructured and professionalized. Not only must deficits with regard to the monetary efficiency of urban water management, as well as those in its knowledge management (e.g. in innovation and investment management), be reduced. Its interaction with spatial and residential development must also be taken more systematically into account. Such a further development, however, faces a complex web of technical and actor-related obstacles which is often not eliminated by spatial planning.

Conclusions

The paper addresses a theoretical and an empirical deficit in the literature on water and regional governance. Hitherto, the interface between technical infrastructure and regional governance has been neglected; their mutual influence has remained largely unobserved. The interlinking conducted here using the example of urban water management assists the recognition of major frontiers of land and water governance. This is of central importance in order, firstly, to enable sustainable collaboration regarding water supply and wastewater disposal that will be able to meet the existing and foreseeable challenges. Secondly, the development of urban regions can be reflected upon in this way and their strategic control can be critically questioned.

The empirical studies demonstrate that it is advisable to organize water management more strongly on a regional level. Major planning and regulation tasks, at least, should be concentrated at the level of urban regions. For a further development of collaboration, firstly, the redistribution of tasks and a stricter alignment of municipal providers with aspects of knowledge management are important starting-points. To this end, individual small-scale institutions need to be strategically linked up in order to facilitate a transfer of know-how. The required changes face strong impediments, however, which complicate internal modernization significantly. Therefore, secondly, external stimuli in the form of incentives for collaboration need to be established politically, particularly by systematic promotion and professional support. Here, opportunity windows would allow the elimination of existing path dependencies. They can occur, for instance, when staff turnovers in management or the comprehensive rehabilitation of technical systems are due. It would be helpful, for example, to integrate needs for rehabilitation local specifically into an urban regional strategy. Collaboration incentives in this context need to accrue from a concerted action of all actors at the different governance levels. The problems of multi-level-governance in turn are impressively recognizable in the lack of interlinking between spatial and infrastructure planning. Technical planning is highly relevant for spatial development in urban regions, but it takes place all too often in the enterprises on a municipal level. In this respect, thirdly, a reorientation of planning seems expedient. Some time ago Healey (2002) observed: 'Spatial planning practices are these days commonly seen as part of the old government, to be displaced and avoided by the new governance forms' (p. 15). This means that planning must form a network among the individual, sectoral actors in urban regions and coordinate their activities.

Also theoretical insights can be gained. There is a bottleneck regarding new governance forms for the mutual renewal of cities or urban regions and their water infrastructures. Nonetheless, in many cases the co-evolution of land use, urban planning and water governance is not systematically traced and conceptualized across disciplines. The case of the Ruhr area demonstrates, however, that when co-evolution is taken into consideration the positive impact of water-related collaboration becomes manifest – not only for water management, but also for regional development. The existing institutional frontiers of improved water governance often stem from technical and historical artefacts. For example the deficit that today's collaboration projects operate in 'wrong' sub-segments can be traced historically. This means that the technical needs of the foundation phase have solidified in regional patterns of thought (even if they have no longer been technically relevant for quite some time). Collaboration projects are strongly aligned with technical needs, and the technical characteristics of infrastructure systems shape their institutional structures. In the course of time, the technical path dependency results in institutional inertness. Major governance schemes thus materialize on the local governance level through complex socio-technical development processes and persistence. This path dependency attains a much greater dimension in technical infrastructures than in social infra-structures, in economic topics or in other areas of regional collaboration.

Even though collaboration regularly fails due to these development processes, such aspects have so far been insufficiently described in governance research. Research on technical infrastructures could significantly enrich the regional governance discourse here. Infrastructure collaboration often reflects regional governance patterns and is therefore an important indicator of regionally specific collaboration arrangements. The analysis of technical infrastructures provides explanatory models for the (non-)functionality of collaboration and thus significantly contributes to the understanding of regional collaboration processes, their bottlenecks and optimization approaches. Thus, in other areas of collaboration what is required is not the creation of new collaboration projects, but rather the restructuring of existing ones, as has been shown for water resource management. Vice versa, the regional governance discourse can contribute to infrastructure research by, for example, expanding the LTS discussion by the spatial dimension that has been missing so far. Thus critical explanatory models of institutional system transformation are provided when information from governance research on the social shaping and regulation of the respective spatial context is included. Furthermore, the stimuli provided by this contribution can also be used for organizational issues of infrastructure systems in an expanded way. The strategic interlinking of sectoral initiatives in an urban region into an interplay of networks and institutionalized structures is not only to be considered a success factor for urban water management, but also constitutes a decisive added advantage if the required meta-sectoral coupling with other areas of sustainable spatial development is to be attained.

Funding

This research was supported by a grant from the Hans Boeckler Foundation.

Notes

1. Only about 530 of the water supply companies are in the five East German states, whereas Bavaria and Baden-Württemberg have about 3700.
2. For an overview of large European water companies, see Hall and Lobina (2012).
3. The 'Blueprint to Safeguard Europe's Water Resources' aims, among other things, at increased efforts towards water conservation, but also at the further reduction of water consumption. The latter increases the problems caused by decreasing consumption described above.
4. To eliminate the micro-pollutants mentioned above (e.g. pharmaceutical products, hormones, lead and zinc), to reduce the amount of nitrogen and to achieve additional levels of hygiene.
5. Collaboration is preferred, for example, to benchmarking, which until now has been implemented much more often in practice.
6. Regarding the dependent variable, it is assumed that there is a high interdependency between water and regional governance, but that present governance forms in household water management are dysfunctional and less integrated regionally, and that they receive no systematic support, e.g. from spatial planning. The causes of this are explained by using different governance patterns and divergent planning practices in different urban regions as the independent variable.
7. For the comprehensive study, see Schmidt (2013).

References

Allouche, J., Finger, M., & Luís-Manso, P. (2008). Water sector evolution scenarios: The case of Europe. *Water Policy, 10*, 221–238. doi:10.2166/wp.2008.149

ATT, BDEW, DBVW, DVGW, DWA, & VKU. (2011). *Profile of the German water sector 2011*. Bonn: wvgw.

Benz, A., & Fürst, D. (2003). Region – 'Regional governance' – Regionalentwicklung [Region – 'Regional governance' – Regional development]. In B. Adamaschek & M. Pröhl (Eds.), *Regionen erfolgreich steuern*. [Controlling regions successfully] (pp. 11–66). Gütersloh: Bertelsmann-Stiftung.

Benz, A. (2001). Vom Stadt-Umland-Verband zu 'regional governance' in Stadtregionen [From urban–hinterland association to 'regional governance' in urban regions]. In Difu (Ed.), *DfK: 2001/II, 40.Vol. Stadt und Region* (pp. 55–71). Berlin: Difu.

Briscoe, J. (1995). *The German water and sewerage sector: How well it works and what this means for developing countries*. Washington, DC: The World Bank.

Coutard, O.Hanley, R. E. & Zimmerman, R. (Eds.). (2005). *The networked cities series. Sustaining urban networks: The social diffusion of large technical systems*. [Roundtable Conference 'Social Sustainability of Technological Networks', New York, April 2001]. London: Routledge.

Danielzyk, R., & Rietzel, R. (2003). Regionalplanung als Motor regionaler Kooperation: Das Beispiel Oderland-Spree [Regional planning as a motor of regional cooperation: The example of Oderland-Spree]. *Informationen zur Raumentwicklung (IzR), 2003* (8/9), 513–521.

Emschergenossenschaft & Lippeverband (2008). *Wo nichts mehr fließt, hilft nur noch pumpen* [Where nothing is flowing the only thing to do is pump]. Essen. Retrieved from http://www. ueberwassergehen.de/content/33/documents/Technik_Pumpwerke.pdf

Federal Ministry for Economic Affairs and Labour. (2005). *Wasserleitfaden* [Water Guideline]. Berlin: PRpetuum GmbH.

Federal Statistical Office. (2009). *Bevölkerung Deutschlands bis 2060* [The population of Germany to 2060]. Wiesbaden: Statistisches Bundesamt.

Furlong, K. (2012). Good water governance without good urban governance? Regulation, service delivery models, and local government. *Environment and Planning A, 44*(11), 2721–2741. doi:10.1068/a44616

Fürst, D. (2003). Steuerung auf regionaler Ebene versus regional governance [Control at the regional level versus regional governance]. *IzR, 2003* (8/9), 441–450.

Fürst, D. (2007). Regional governance. In A. Benz, S. Lütz, U. Schimank, & G. Simonis (Eds.), *Handbuch Governance* [Governance manual] (pp. 353–365). Wiesbaden: VS.

German Association of Energy and Water Industries (2011): *Wasserfakten im Überblick* [An overview of water facts], Berlin. Retrieved from http://www.bdew.de/internet.nsf/id/ DE_Wasserfakten_im_ueberblick/$file/11%2001%2012%20Wasserfakten%20im%20%20% C3%9Cberblick%20-%20%20Januar%202011.pdf

German Bundestag – 16th legislative period (2006). *Bericht der Bundesregierung zur Modernisierungsstrategie für die deutsche Wasserwirtschaft und für ein stärkeres internationales Engagement der deutschen Wasserwirtschaft* [Report by the German government on the modernization strategy for the German water industry and for a stronger international involvement by the German water industry]. Drucksache 16/1094. Berlin.

Graham, S., & Marvin, S. (2008). *Splintering urbanism: Networked infrastructures, technological mobilities and the urban condition*. London/New York: Routledge.

Green, C., & Anton, B. (2012). Why is Germany 30 years ahead of England? *International Journal of Water, 6*(3/4), 195–214. doi:10.1504/IJW.2012.049496

Grigg, N. S. (2011). Water governance: From ideals to effective strategies. *Water International, 36* (7), 799–811. doi:10.1080/02508060.2011.617671

Hall, D., & Lobina, E. (2012). *Water companies and trends in Europe 2012*. Brussels: EPSU. Retrieved from http://www.epsu.org/IMG/pdf/2012_Water_companies-EWCS.pdf

Healey, P. (2002). Spatial planning as a mediator for regional governance. In D. Fürst & J. Knieling (Eds.), *Studies in spatial development. Euro Conference 19.–21. April 2001, Hannover* (pp. 13–25). Hannover: ARL.

Heinelt, H., & Zimmermann, K. (2011). How can we explain diversity in metropolitan governance within a country?' Some reflections on eecent developments in germany. *International Journal of Urban and Regional Research, 35*(6), 1175–1192. doi:10.1111/j.1468-2427.2010.00989.x

Heyen, N., Kaika, M., & Swyngedouw, E. (2006). Urban political ecology: Politicizing the production of urban natures. In N. C. Heynen, M. Kaika, & E. Swyngedouw (Eds.), *In the nature of cities* (pp. 1–20). London: Routledge.

Hodson, M., & Marvin, S. (2007). Cities mediating technological transitions: The adaptability of infrastructure and infrastructures of adaptability? In H. S. Geyer (Ed.), *International handbook of urban policy* (pp. 240–258). Cheltenham: Elgar.

Hodson, M., & Marvin, S. (2009). Cities mediating technological transitions: Understanding visions, intermediation and consequences. *Technology Analysis & Strategic Management, 21*(4), 515–534. doi:10.1080/09537320902819213

Holländer, R., Fälsch, M., Geyler, S., & Lautenschläger, S. (2009). *Trinkwasserpreise in Deutschland* [Drinking water prices in Germany]. Leipzig. Retrieved in shortened form from http://www.vku.de/fileadmin/get/?14512/pub_Kernaussagen_Holl%C3%A4nder_II_091101.pdf

Holländer, R., Lautenschläger, S., Rüger, J., & Fälsch, M. (2013). *Abwasserentgelte in Deutschland* [Wastewater fees in Germany]. Berlin/Munich: Sigillum.

Kaika, M., & Swyngedouw, E. (2000). Fetishizing the modern city. *International Journal of Urban and Regional Research, 24*(1), 120–138. doi:10.1111/1468-2427.00239

Kilper, H. (2007). Innovation als Herausforderung: Strategien in Zeiten von urban und regional governance [Innovation as a challenge: Strategies in times of urban and regional governance]. In

S. Kelp-Siekmann, P. Potz, & H. Sinning (Eds.), *Innovation und regionale Kooperation* [Innovation and regional collaboration] (pp. 57–68). Dortmund: IfR.

Knieling, J. (2003). Kooperative Regionalplanung und regional governance [Collaborative regional planning and regional governance]. *IzR, 2003* (8/9), 463–478.

Lackowska, M. (2011). Metropolitan governance. In H. Heinelt, E. Razin, & K. Zimmermann (Eds.), *Interdisciplinary urban research: Metropolitan governance* (pp. 21–31). Frankfurt/Main: Campus.

McFarlane, C., & Rutherford, J. (2008). Political infrastructures: governing and experiencing the fabric of the city. *International Journal of Urban and Regional Research, 32*(2), 363–374. doi:10.1111/j.1468-2427.2008.00792.x

Mehta, L., Marshall, F., Movik, S., Stirling, A., Shah, E., Smith, A., & Thompson, J. (2007). *Liquid Dynamics: challenges for sustainability in water and sanitation*. Retrieved from http://www.steps-centre.org/PDFs/final_steps_water.pdf

Monstadt, J. (2004). *Die Modernisierung der Stromversorgung: Regionale Energie- und Klimapolitik im Liberalisierungs- und Privatisierungsprozess* [The modernization of electricity supply: Regional energy and climate policies in the liberalization and privatization process]. Wiesbaden: VS.

Monstadt, J. (2007). Großtechnische Systeme der Infrastrukturversorgung [Large technical systems in infrastructure supply]. In D. Gust (Ed.), *Wandel der Stromversorgung und räumliche Politik* [Changes in electricity supply and spatial policies] (pp. 7–34). Hannover: ARL.

Monstadt, J. (2009). Conceptualizing the political ecology of urban infrastructures: Insights from technology and urban studies. *Environment and Planning A, 41*(8), 1924–1942. doi:10.1068/a4145

Monstadt, J., Schmidt, M., & Wilts, H. (2012). Regionale Zusammenarbeit in der Ver- und Entsorgung des Rhein-Main-Gebiets [Regional collaboration in supply and disposal services in the Rhine-Main area]. In J. Monstadt, T. Robischon, B. Schönig, & K. Zimmermann (Eds.), *Die diskutierte Region – Probleme und Planungsansätze der Metropolregion Rhein-Main* [The discussed region – problems and planning approaches in the Rhine-Main metropolitan region] (pp. 185–210). Frankfurt am Main: Campus.

Pütz, M. (2004). *Regional Governance: Theoretisch-konzeptionelle Grundlagen und eine Analyse nachhaltiger Siedlungsentwicklung in der Metropolregion München* [Regional Governance: Theoretical and Conceptual Foundations and an analysis of sustainable residential development in the Metropolitan Region of Munich]). München: Oekom.

Reuther, J., & Schreiber, U. (2009). Alles im Fluss: Die Nutzung von Ruhr, Emscher und Lippe [Everything in flow: The utilization of Ruhr, Emscher and Lippe]. In A. Prossek, H. Schneider, B. Wetterau, H. A. Wessel, & D. Wiktorin (Eds.), *Atlas der Metropole Ruhr* [Atlas of the Ruhr metropole] (pp. 192–193). Köln: Emons.

Schmidt, M. (2013). *Regional Governance und Infrastruktur – Kooperationen in der Wasserver- und Abwasserentsorgung am Beispiel der Stadtregionen Frankfurt/M, Berlin und Ruhr* [Regional Governance and Infrastructure – Collaboration in water supply and wastewater disposal in the urban regions of Frankfurt/M, Berlin and the Ruhr]. Detmold: Rohn.

State Parliament of Hesse (2010). Gesetzentwurf der Fraktionen der CDU und FDP für ein Gesetz über die Metropolregion Frankfurt/Rhein-Main und zur Änderung anderer Rechtsvorschriften vom 30.08.2010 [Draft of 30.08.2010 by the CDU and FDP Group for a law on the metropolitan region Frankfurt/Rhine-Main and changes in other legislation]. Drucksache 18/2733, Wiesbaden.

Störmer, E., Truffer, B., Dominguez, D., Gujer, W., Herlyn, A., Hiessl, H., ... Ruef, A. (2009). The exploratory analysis of trade-offs in strategic planning: Lessons from regional infrastructure foresight. *Technological Forecasting & Social Change, 76*(9), 1150–1162. doi:10.1016/j.techfore.2009.07.008

Truffer, B., Störmer, E., Maurer, M., & Ruef, A. (2010). Local strategic planning processes and sustainability transitions in infrastructure sectors. *Environmental Policy and Governance, 20*(4), 258–269. doi:10.1002/eet.550

Wackerbauer, J. (2007). Regulation and privatisation of the public water supply in England, France and Germany. *Competition and Regulation in Network Industries, 8*(2), 101–117.

Wackerbauer, J. (2009). *The Water Sector in Germany*. Working paper CIRIEC No. 2009/11. Liège.

Wackerbauer, J. (n.d). *Public or private water management: Experience from different European countries*. Munich. Retrieved from http://ksh.fgg.uni-lj.si/bled2008/cd_2008/04_Water%20management/011_Wackerbauer.pdf

Zschille, M., & Walter, M. (2011). The performance of German water utilities: A (semi)-parametric analysis (discussion papers no. 1118). Berlin.

Managing urban riverscapes: towards a cultural perspective of land and water governance

Meike Levin-Keitel

Institute of Environmental Planning, Leibniz University of Hanover, Hanover, Germany

Urban riverscapes are facing diverse demands concerning riparian uses, ecological, economic and social functionalities, and aesthetic questions. One of the main challenges today is the implementation of an integrative perspective on riverscapes to overcome the horizontal frontiers of traditional water management (water governance) and urban planning (land governance). Led by the theoretical framework of planning culture, the article shows the different rationalities and governance approaches from a cultural perspective. Finally, two quite different local planning cultures are outlined to illustrate the cultural variety with which the challenges of sustainable urban riverscapes are managed.

Focusing on urban riverscapes

Rivers have always played a crucial role in human activities, e.g. the founding of settlements on riverbanks, the use of temporarily flooded areas for agricultural food production, or their use as strategic barriers of defence. On rivers, goods for trade were transported, the river water was necessary for the development of different techniques of traditional handicraft, and it simultaneously served as a fresh water supply and wastewater disposal. Nowadays, focusing on urban riverscapes includes being confronted by some of these issues (albeit slightly modified), as well as the multiple additional demands on the river water itself (in terms of quality and quantity) and on its riparian land uses and their diverse influences (in terms of the cycle of substances, interdependence between these uses and the river ecosystem, their spatial impact, the consequences of societal changes, etc. (Levin-Keitel, 2014)). Especially in urban areas, where space by the river is limited and numerous diverging interests are articulated, these demands and peculiarities are gaining a controversial dimension.

Multiple policy perspectives and their integration

Even though the city level is neither typical for nor representative of river management in general (such as the scale of international river associations, river basins or regional catchment areas), the questions of integration and integrated implementation occur at the local level, often in a project-based approach. Accordingly, in the last decade urban riverscapes have re-emerged as one of the central issues of European urban planning and development. Due to different driving forces, urban riverscapes come to the fore of diverse stakeholders, demands and functionalities, and the promotion of various laws, directives, programmes and policies. This includes, in particular, European guidelines such as the European Union Water Framework Directive, the current discussion about urban climate change adaptations, and former industrial sites in inner cities being redeveloped. These diverse interests mean that the idea of an integrative perspective represents a significant challenge, which consequently raises considerations about aspects of traditional water management, such as water quality and flood management. Furthermore, only respecting and considering multiple demands, including economic interests, ecological aspects, historical significance, sociological conditions or infrastructural perspectives, allows one to develop sustainable urban riverscapes. Certainly, one can identify a number of policy sectors articulating river purposes in general (Figure 1); but as local circumstances play an important role, the art of management lies in juggling the appropriate purposes without underestimating their local interdependence.

This shows a first insight into the complexity and enormous challenges that have to be met in managing urban riverscapes. Each policy sector has its eligibility, each sector formulates essential needs and demands, and an integrative perspective indispensably has to understand and tolerate these specific socially constructed views by accepting riverscapes as product of many separately made decisions. Or, as Biswas (2004) wrote concerning integrated water resource management (IWRM)[1]: 'water problems have become multi-dimensional, multi-sectoral, and multi-regional and filled with multi-interests, multi-agendas, and multi-causes, and which can be resolved only through a proper multi-institutional and multi-stakeholder coordination' (p. 249).

This article focuses on the indispensable integrative part of managing urban riverscapes. This involves the different facets of water and its governance from a land perspective as well as from a water point of view. Being far off just a 'management problem', the challenge of an integrative approach lies in overcoming diverse rationalities

Figure 1.   Multiple policy sectors articulating river purposes.

and habits, combining different legal, formal and informal frame conditions, creating win–win situations whenever possible and/or establishing so-called intercultural learning processes to overcome not only horizontal cultural frontiers. Core questions are as follows:

- What is the role of culture (like norms and values) in integrated planning processes of urban riverscapes?
- How can these dynamic cultural complexes be defined and analysed in terms of key aspects of organizational cultures?

Empirical studies and methodological approach

This article builds up on two empirical case studies being part of a major PhD project treating cultural perspectives of integrated planning processes in theory and practice. The two case studies: Ratisbon (Regensburg) and Nuremberg (Nürnberg), are situated in Bavaria in southern Germany.

Ratisbon has 138,000 inhabitants (Stadt Regensburg 2012, p. 10) and is well known for its medieval inner city and picturesque riverscape. Nearly all parts of the inner city have been a UNESCO World Heritage site since 2006 (United Nations Educational, Scientific and Cultural Organisation (UNESCO), 2014) and an authentic Bavarian style of living is presented to the numerous tourists from all over the world (nearly 1 million nights per year; Stadt Regensburg 2012a, p. 9). Three rivers characterize the city substantially: the Danube, the Regen and the Naab. While the Naab flows into the Danube before crossing the urban realm, the Regen joins it from north. The Danube itself is split up into three branches in the city area, one navigable canal and two more or less naturally river branches. As result, three islands and many (medieval) bridges are part of the city's image. In general, Ratisbon is quite vulnerable to hazards of flooding, the major reason to rethink the urban riverscapes.

Nuremberg is characterized by nearly 500,000 inhabitants and a population density of 2655 inhabitants per km^2 (Stadt Nürnberg 2012, n.p.). Sandstone, small streets and houses as well as few open spaces dominate the inner city, and apart from some interesting touristic destinations, the demands of the local inhabitants are on the agenda (e.g. qualitative open spaces, transport and mobility). The river Pegnitz runs through the city feeding a huge sea in the inner city, the Wöhrder See, and creating inhabited islands in the river itself. Apart from the sea situation, water is not very visible in the inner city. In the 1970s, the management of Nuremberg's floods found technical solutions in forms of canals and banished all water from the open space. This is why nowadays water must be rediscovered as element in the inner city.

A comparison between the two case studies in Ratisbon and Nuremberg is limited neither to their urban morphology nor to their later history in managing riverscapes, nor to their intentions to turn to their riverscapes nowadays. Here the planning process of an integrative perspective on urban riverscapes has focused on drawing conclusions on the cultural perspective of integration processes as well as the organizational cultures of the involved actors. Two integrated urban development plans have been extensively researched: in Ratisbon the riverscape concept (WWA & Stadt Regensburg, 2010) and in Nuremberg the urban development concept water (Stadt Nürnberg, 2012b), in order to work out the actor's opinions and rationalities, interpretations of socially constructed realities, sector-based instruments and their implementation, as well as their specific language and characteristics. Methodologically, these two very complex planning processes where analysed by a diverse mix of methods of structured expert interviews, participating observation, and local document research, from where the information and data are taken. On the one hand, Ratisbon and Nuremberg have been chosen as case studies because of their similar point of departure, for instance their identical legal bases, the same organizational structures of the involved institutions and the same formal planning system. On the other hand, both cities show differences in how they handle and manage the urban riverscapes, beginning with the perception of water in the inner cities, the leadership of the integrated planning process (the water management agency in Ratisbon and the local planning authority in Nuremberg), and the common ground from which they started. This was intended to make the cultural differences visible that are not simply part of another planning system (as in European comparative research for instance) or differing legal conditions (like in different German states).

So after having briefly introduced urban riverscapes and having pointed out the high complexities and interdependencies of their management, this article calls on the cultural perspective of land and water governance. In the first step, the concept of governance as an actor-centred view is represented and applied to urban riverscapes, revealing the limits of this concept by considering perceptions and interpretations of problems and solutions. To overcome these rationalities, perspectives and aforementioned horizontal frontiers, the theoretical framework of organizational culture is represented in a second step. Using culture as an analytical framework, the integrated planning processes of urban riverscapes are presented by their visible artefacts, organizational culture and planning context. Following this, different cultural imprints are illustrated, with special focus on two different modes of governance: water governance and land governance. The article ends with a plea for a cultural perspective on integration processes and the added value it brings.

Just a question of governance – an actor-centred view

Taking into account the multi-thematic approaches of urban riverscapes, the need for a certain kind of management becomes evident. As the local government or even the public sector is not the only controlling actor when urban riverscapes are considered, other ways of steering and managing such sites come to the fore. Generally, governance theory is divided into three different currents: the analytical purpose of steering and decision-making, the normative purpose of good governance, and the descriptive purpose where the content plays a crucial role (Benz, Lütz, Schimank, & Simonis, 2007). Using governance theory as an analytical framework to describe contemporary planning processes also means considering not only the government but also a whole ensemble of actors, institutions and management (Benz et al., 2007; DiGaetano & Strom, 2003; Fürst, 2001; Nuissl & Heinrichs, 2011). In particular, environmental governance was meant to be the solution for all issues and measures for avoiding harmful effects on the environment (Driessen, Dieperink, van Laerhoven, Runhaar, & Vermeulen, 2012). In the urban riverscapes, possible conflicts become visible: multiple actors at all levels with specific rationalities are trying to reach consensus on land and water issues in a particularly small place: the urban riverscape. Table 1 outlines the main actors in all different policy fields with their kind of rationality in the sense of an operating logic of their means and ends as well as their anchoring on different spatial and administrative levels. The contents are based on the findings of the two case studies of Ratisbon and Nuremberg. Due to the high degree of contextuality in urban planning and development,

Table 1. Multi-actor, multilevel governance of urban riverscapes.

	Actor	Rationality	Levels	Instruments
Public sector	Water management agencies	Water supply, water quality, environmental renaturation; Implementing sectoral planning	European National State Regional Local	Own laws, directives, plans, programmes and projects
	Local planning authority	Urban planning and development, integrated issues; Community planning sovereignty	Local	Own laws, land-use plans and construction plans
	Environmental authority	Environmental renaturation, biodiversity, ecosystem approach (flora and fauna); Implementing sectoral planning	State Regional Local	Own laws, environmental/landscape plans, programmes and projects
	Transport planning authorities	Infrastructure and mobility, riverscapes as linear connections in the dense city centre and as obstacles to overcome; Implementing sectoral planning	European National State Regional Local	Own laws, directives, plans, programmes and projects
	City council	Political decisions and strategies oriented in legislative terms; policy positioning	Local (depending on the national level and state level)	Policy documents and decisions
	Historic conservation authorities	Protection of historical, natural and cultural monuments; Sectoral planning	UNESCO at European level National State Regional Local	Comprehensive laws, directives, plans, programmes and projects
	Tourism	Touristic economy and industries; Profit-oriented optimization	Regional Local	Investments, lobbying, economies
Private sector	Real estate industry	Waterfront development, rediscovery of urban waters for economic uses, regional identity	Regional Local	Investments and financing, markets, lobbying
	Port industry	Rediscovery of urban waters for economic uses	National Regional Local	Investments and financing, markets, lobbying
	Citizens and interest groups	Various rationalities, e.g. local residents, associations or initiatives for fishing, historic conservation, accessibility of riverbanks, etc.	(National) Regional Local	Lobbying, land owners, democratic impact

Source: Author.

this overview just helps to illustrate the complexity of such governance approaches and is no claim to universality.

Regarding this diversity of stakeholders, their specific rationalities on how they perceive the urban riverscape and the objectives they come along with, a highly rational way of steering the development of urban rivers seems to be desperate. Especially when it comes to the point where neither a single institution nor a single rationality is powerful enough to be able to predominate over other stakeholders, institutions or rationalities. Moreover, it must be held that there is not one institution coordinating riverscapes at any level; in fact, it is more or less like an overlapping mosaic of institutions and organizations' competence.[2] How then is it possible to steer these different views and actors? What does integration in this very complex aggregation of actors, power and discourses mean? Making integration a subject of discussion means not remaining on purely normative projections and notions of what seems to be 'right' or 'wrong'. In fact, it is rather a question of tapping into the available potential of integrating different demands at the urban riverside. The aim is not a comprehensive plan including every single probability proven by facts and figures, but an integrative planning approach, considering uncertain conditions and an at present unpredictable future (Mitchell, 2005). Integration can be seen as a creative process for structuring a mosaic of different layers for the optimum interplay, so the first point is to accept this very complex and highly polyrational steering process. Even if all available facts and every actor involved are integrated, the result of planning processes in urban riverscapes is not predictable in all details (Billé, 2008). Consequently, it is not surprising that the management of diverse different rationalities on various levels is highly context-bound. As Rittel and Webber (1973) stated in their theory of wicked problems in spatial planning, policy problems cannot be definitely described because they refer to public goods with no objective definition of equity, so one cannot talk about an optimum solution. So, what is called a solution only fits in one specific context and can become obsolete immediately. The incontestable aspect is that planning processes are highly context-bound to the extent that they call for individual solutions and open process designs. This explains why urban riverscapes look different in New York and Berlin, why the process design of developing urban riverscapes occurs differently in Paris and Hamburg; or, as Driessen et al. (2012) claim, the 'extent of multi-actor, multi-level governance determines variation in the perception of problems and solutions' (p. 145).

The role of culture in an actor-centred approach

The question arises of how these context-bound differences in managing urban riverscapes can be analysed and what this means for their development. In a phenomenological approach, however, even if one compares the planning process of two urban riverscapes in two cities like Nuremberg and Ratisbon – with the same institutional organizations, identical legal conditions and the same groups of actors involved – significant differences can be noted. At this point, a cultural dimension of planning processes and an awareness of culture as a matter of fact in planning processes becomes apparent.

Theoretical derivation of a cultural perspective on integration processes

From the perspective of planning theory, this cultural turn can be considered as an approach to gaining a deeper understanding of planning processes. The discourse of planning culture as a whole is derived from two different aspects: (1) by recognizing the variety and the differences of planning institutions and practices (Friedmann, 2005) or (2) spatial planning as a cultural act itself (Nuissl, 2008). After the often claimed 'communicative turn' (Healey, 1996) in planning theory, a multi-actor, multilevel governance approach represented a kind of shift from a highly rational practice of management to a planning process determined by communication, cooperation and participation. With the communicative turn, the focus of analyses was set on the actors of a system, an actor-centred approach (Fürst, 2007; Nuissl & Heinrichs, 2011). However, a general gap between the theoretical interpretations and real planning practice can still be stated. Dependent on an authority's behaviour and cooperation, and willingness to communicate, the involvements in waterside projects differ. Therefore it is not all about institutional arrangement and the organizations involved, but another important factor is outlined: the cultural orientation of the involved actors and the planning culture they establish and maintain (Knieling & Othengrafen, 2009; Othengrafen, 2012). Building up on the definition of culture by Gullestrup (2009) the concept of culture can be transferred to actor-centred integration processes:

> Culture is the world conception and the values, moral norms and actual behaviour – as well as material and immaterial results thereof – which people take over from a past generation; and which make them different in various ways from people belonging to other cultures. (p. 4)

So, the cultural perspective is about a society's cultural imprint, meaning the cultural behaviour of all involved actors, different cultural techniques, cultural attitudes and their interaction with each other. In the literature, planning culture is often used with terms like cultural framing (Ernste, 2012), as culturized planning practice (Othengrafen 2010) or planning culture as an underlying concept expressing the practice of planners (Reimer & Blotevogel, 2012, p. 14). Consequently, 'Planning systems and planning practice are part of the culture of a society' (Nuissl, 2008) and the other way round, planning processes produce, develop and maintain the culture of a society by the way in which cities are designed, developed and influenced. Nevertheless, culture is ambiguous, complex to analyse and to identify if it is not taken as a normative ideal in the sense of a better/ good planning culture. The aim of this article is not about identifying how planning should be or should theoretically come out. This implies not sticking to one truth but rather explicitly emphasizing the perception and interpretation of single actors and groups of actors.

Planning culture as analytical framework for urban riverscapes

How can these characteristics lead to an analytical framework for the analysis of urban riverscapes? The aim is to understand the developing process of integrated riverscape management, especially the different perspectives on the river, ranging from sectoral logics of action and self-conceptions to common orientations and the shared learning process (innovation ability). The following analytical framework builds on a theoretical planning culture model of Levin-Keitel and Sondermann (2014). These authors identify a core societal context, as planning is part of a society's culture, framing all integration processes and playing a key role in the comparison of planning cultures in international or European contexts. Apart from this omnipresent societal dimension, three main analytical levels representing cultural elements are identified:

- Visible artefacts represent the products and results of urban planning and development, implying spatial plans and concepts as well as their specific implementation in the form of spatial structures and functions.
- Planning context signifies the setting in which planning processes take place, including the main structures and frames of the planning conditions. Here the structures of the institutional and legal system, as well as formal and informal rules and procedures, stand in the focus of the reflection.
- Organizational cultures denotes an organization's values and norms, its specific processes of learning and socialization as inter-individual patterns, its self-conception and game rules. These cultural elements can be stated amongst a single group of actors and additionally between different groups of actors, making it even more complex and interesting.

First, visible artefacts include in particular all spatial plans and concepts, as they can be seen as a specific language where present values and norms are manifested. Furthermore, the content of these plans and concepts mirrors the contemporary spirit of spatial planning by identifying the most important challenges and the field of action, based on (cultural) interpretations of what is important and what is not (Othengrafen, 2012). In the last 20 years, many cities have rediscovered their urban riverscapes, and water and rivers have played a crucial role in integrated city development plans again after having been ignored for decades. Nowadays the implementation of these ideas can be re-found in the urban space. Therefore, secondly, contemporary spatial structures and functions in the inner city have to be read as cultural artefacts as well. They are so to speak cultural witnesses in stone: on the one hand, these structures and patterns can be seen as an expression of the human-made plans and development of former times; and, on the other hand, they are the starting point for every (spatial) activity now and in future times. Ideas, plans and concepts become reality – they become de facto existing structures and patterns and symbolize an expression of culture. These visible artefacts determine to a high degree contemporary planning culture, including which demands have to be integrated – whether flood protection is needed, how citizens can be involved, etc. – and, furthermore, questions of land use – whether to place settlements in a flood-risk area, how to combine or to separate different land uses, etc.

The planning context is about the planning conditions, such as formal and informal rules and the institutional planning system. It covers the specific framing conditions of spatial planning, unlike the visible artefacts that are a result of planning endeavours as well as of other non-planning-related developments. This planning context implies legal conditions such as the approval procedures of constructions in the flood-risk areas or sectoral legal demands, administrative and organizational rules such as hierarchical positions, and the actors who are to be involved in legally binding processes. Informal and formal rules and arrangements play an important role in planning processes as well, as they allow enough flexibility for actions and routines. The planning context is the central point of departure for actors in the planning processes and their perceptions, interpretations, judgments and actions. For a further analysis of this part of planning culture, compare the already established system-related analysis in spatial planning, where numerous elements of scope are already discussed (Ernste, 2012; Othengrafen, 2012; Reimer & Blotevogel, 2012).

Organizational cultures involve all individuals and groups acting on the local, regional or national level. This includes the local planning authorities and public services, citizen stakeholder groups or economic actors. Focusing on the specific constellations of and interactions between the actors (Fürst, 2007; Moss, 2009; Nuissl & Heinrichs, 2011), this part of a planning culture is about individual perceptions, institutionalized interpretations and their resulting actions. The consideration of all involved stakeholders in planning processes is by no means new or innovative, but represented by Scharpf (2000) and implemented within the governance analysis. The new and specific cultural aspect of these interactions and constellations is to regard different actor groups as organizations with an explicit organizational culture. The term 'organization' is used here in a very broad sense – an organization is a group of actors more or less institutionalized, ranging from the local planning authority or water management agencies to single-purpose citizens' initiatives or clubs and associations. Organizational cultures can be defined as the result of processes of learning and socialization processes under a certain umbrella of an organization. These are based on invisible concepts such as values and norms, and lead to:

- the formation of common patterns of orientation ending up in central premises of action;
- the development of shared informal rules; and
- the characterization of the self-conception of an organization (Schein, 2003).

From an external perspective, organizational cultures are noticed by visible perceptas, namely the visible artefacts of the analytical model presented above. The concept of the organizational culture is based on the theoretical cultural approach of Schein (2003). This concept was implemented by Faust (2002) on public services and by Othengrafen (2012) on spatial planning. As public services are responsible for spatial planning at diverse levels of planning, this concept seems appropriate for the analysis of the cultural dimension of spatial planning.

Cultural perspective of land and water governance

The aforementioned analytical framework is now implemented in the two case studies on the trail of local planning cultures, with special interest in the organizational cultures of water management agencies and local planning authorities in Ratisbon and Nuremberg. The following information and data are mostly based on the expert interviews made on site (if not cited differently).

Visible artefacts contain the spatial plans and concepts within the two cities of Ratisbon and Nuremberg. In both cities one can observe a high number of formal and informal plans and programmes, concepts and politics in different coalitions of actors, ranging from area-based integrated development plans (e.g. for the western city, for urban redevelopment) to thematic or sectoral concepts for the whole urban area (e.g. a mobility plan or programmes for cultural heritage). However, both cities are very vivid concerning spatial development, being confronted with several pressing challenges. Additionally, both cities' societies are discussing these concepts and plans intensively in local newspapers, discussion forums or participation activities. Furthermore, the structure and the functions of the inner cities, as a second part of the category of visible artefacts, are comparable as well. Both cities are characterized by a medieval inner city, in the outer scope of flood areas. This includes a rather densely built environment with a potential lack of open space already. In both cases, different demands of restructuring the urban riverscape meet the challenge of an already compactly built city structure. The rivers – in Ratisbon the Danube, Naab and Regen, and in Nuremberg the Pegnitz – flow through the inner city, creating inhabited islands as well as zones of natural protection. Even if Ratisbon is, for instance, potentially at danger of floods (because of its spatial exposure) and Nuremberg is more affected by droughts (and invisible water in the city area), the riparian uses and the complex planning processes for restructuring and managing the riverscapes are comparable.

Apart from the detailed table of multi-actor, multilevel governance (Table 1), some supplementary information has to be added. For the two chosen case studies, the planning context is nearly identical, as both cities are situated in Europe (European Directives: FRD, WFD), in Germany (national framework: WHG), and in Bavaria (the same federal state meaning the same water legislation: BayWG) with the same administrative and organizational background of decision-making. The planning context of the case studies is mainly predetermined by two actors and their legally binding instruments: the water management agencies and the local (urban) planning authorities. Both of them do have an integrative approach in space and time, and both are part of the public services acting in line with their proper water purpose legislations, which have binding consequences for the other governmental services and the private sector (Von Haaren & Moss, 2011; Wiering & Immink, 2006).

Water management agencies are, by WHG §38(3), in charge of the river itself and the sufficiently broad stretches of the riparian uses. Here the water management agencies (in representation of the Bavarian Free State) work as contact persons on a regional scale, not only within the city boundaries. Their instruments (river basin management plans etc., WHG §83) are binding; they plan and implement with their own financial and human resources. For construction they need a permit from the city. Local (urban) planning authorities are part of the city administration and very close to politics, as their leader is elected politically (city council). Their aim is city development and a balance of ecological, economic and social demands within the city area. Their perspective on the riverscape is characterized by land uses, riparian uses or the obstacle of the river in the city area. Their instruments are binding for public services (land-use plans, BauGB §§5–7) and for all citizens (construction plans, BauGB §§8–10). The urban planning authorities plan, but the implementation lies in the hands of private investors or sectoral planning agencies. They are not able to implement projects on the river by themselves, as they do not possess their own financial resources. Therefore, even if there are some regulations within the management of riverscapes that direct them to fulfil the demanded cooperation between the water management agencies and the urban planning authorities, there is still a lot of freedom of interpretation as to how this cooperation can be filled with life. Table 2 identifies the two different rationalities water management agencies and local planning authorities display by their objectives, main interests, intentions, planning focus and temporal and financial resources. Rationalities are hereby seen as rational conclusions of their mission.

In the following, the organizational cultures of water management agencies, as well as urban planning authorities in Bavaria, are drawn as they were analysed from the two case studies. Of course, these following arrangements are not representative for all water management agencies and urban planning authorities; furthermore, they give an idea about what the organizational culture is about and to where the specific cultural imprint leads. Both organizational cultures are described, analysed and abstracted by four main categories, evolving from the empirical analyses:

- important traditions and contemporary orientations;
- concepts of democracy and justice;
- effectiveness of their outputs; and
- self-conception and ability to innovate.

Table 2. Comparison of governance of water management agencies and urban planning authorities.

	Water management agencies	Urban planning authorities
Governance mode	Water governance	Land governance
Objective	Water resources (multifunctional)	Land use (multidisciplinary)
Main interest	Safety of the city	Urban development and design
Intention	Good ecological quality (or potential of water bodies)	Principles and aims of regional policy
Planning focus	Single projects in regional scales, spatial vision of the river	Comprehensive and strategic (whole city-area)
Time schedule	Standardized European Union-wide time schedule	Open, different spatiotemporal process
Resources	Comparatively good financial and personnel resources	Comparatively less financial and personnel resources

Source: Author.

The water management agencies of the two case studies do have an implementation-oriented governance approach, as already mentioned above. This is characterized by the following:

- Their interactions are based on federal water law with the aim of ensuring water supply, to minimize flood risk and to develop nature-like environments (BayWG §§22–23). In former times, the water management agencies were much more technically oriented, being specialists in the construction and measurement of rivers and seas. This basis on natural science techniques to survey, measure and control water in general has undergone a paradigm shift in the last 30 years (Driessen et al., 2012; Gleick, 2000; Hartmann, 2010; Molle, 2008). The keywords of this development are the uncontrollability of nature and the failure of technological solutions, unpredictable changes, such as climate change, demographic changes, etc. Contemporary orientations in the water sector are characterized by a shift from purely technological and natural scientific solutions to a coordination of societal needs. The questions are not only what is technically feasible and in how far it is the best solution, but also what kind of land use is desired, how flood security can look under these circumstances, and how society wants to live. In other words, values and norms are threatened as well as technical solutions. Here the second point of the analyses is already introduced. Of course, as a public service, the water management agencies are integrated in the democratic machinery.
- They also belong to the federal administration, to an elected ministry, legitimated by politics, laws and assessments. Furthermore, they are obliged to organize public participation. Nevertheless the question of justice remains. How much open space by the river has to be provided to the general public? Do water management agencies take care of the common goods? And, in terms of flood-risk management, how are citizens' demands to be integrated? Is it just that a couple of citizens refuse to construct a protective wall in front of their house, but at the same time call for help in situations of flood catastrophes (putting firemen in danger as well)? The clash of security needs and participation doubtlessly demands a balancing act and remains one of the central issues of water management agencies in future.
- In terms of effectiveness, water management agencies achieve rather good outputs, especially as the achieved quality becomes obvious: water supply is guaranteed

whenever water is needed, flood security is proven by every flood and so is the quality of the built and organizational measures. With regard to renaturation, results can be analysed by red-listed animals or plants or other natural science proofs.

- As sectoral planning agencies, they possess high self-confidence, further strengthened by numerous developments and versatile competence on all levels. In the past, the water management agencies already played a crucial role: water supply and water disposal as basic needs in the cities, making rivers and seas navigable for economic uses, flood security for densely populated and valuable regions, co-determination of the image of the cities, etc. This importance was reinforced by additional tasks, e.g. by the water framework directive and national/federal implementation. The reorganization of the water management agencies led to a very modern, open-minded and innovation-friendly authority, on the one hand acting in area-based teams on all levels and scales and, on the other hand, in thematic expert teams transferring new knowledge and experiences directly in all levels. Another asset is the medium size of the water management agencies (about 2000 people), which supports the transfer of ideas, visions, as well as problems, challenges and threats.

Nevertheless, in both agencies, after a time of warming up and getting to know each other, the interview partners were able to draw a greater vision of the rivers' development and the urban water development, which has never been written down before, but which has been created by consensus in the specific water management agency. It is, so to say, a part of the invisible underlying core assumptions.

In comparison with this, the organizational culture of the urban planning authorities can be portrayed by the logic of land governance:

- They operate from a territorial perspective, following the riparian uses towards the water. As urban planning and development aim for an integrative perspective, diverse demands have been collected, evaluated and a decision on the most expedient solution has been made, so one can state that all relevant aspects are under consideration. Yet, as continuous changes (climate change, demographic change, structural change) and modifications in the (spatial) needs and demands of a society evolve permanently, urban planning always remains in a discursive approach between many possible solutions. Apart from the legally binding instruments, the power of spatial planning in general is based on persuasive instruments, by arguing and convincing the involved actors to implement what planners have planned before (Hutter, 2007).
- Concerning the concept of democracy and justice, unquestionably the urban planning authority is part of the cities' administration, led by local politicians (city council), with its proper legal basis and binding instruments. So, the democratic legitimation is given by the elected city council and key decisions were prepared by the administration, but still the decision lies with the responsibility of politics. Here, the two case studies differ obviously. While in Ratisbon the main issue of the riverscape management, flood protection, is not a political debate in the urban society, in Nuremberg the accessibility of the river (as the main issue here) is politically implicated. The situation gets even more complex by recognizing that the urban planning authority is not only one administrative body in the city administration, but (differing from city to city) consists of different departments (e.g. the urban development department, the urban planning department, the

environmental planning department, the construction department), which are in most cases led by representatives of different political parties. In this sense, already within one organization, a far-reaching negotiation process has to take place, and is not always in consensus about how the urban environment (and especially the urban riverscapes) is to be managed, designed or treated.

- This is part of an explanation as to why the effectiveness of their outcomes is hard to define, or nearly impossible to judge. First of all, the question arises: effective for what? Is contemporary land use effective for future generations (regarding the theory of sustainability)? Is the use of the ground effective or not? In general, spatial planning has to deal with this problem of evaluation. There are no real accurate criteria with which to judge the outcomes of spatial planning (Rittel & Webber, 1973). For instance, is a consensus in planning processes worth more than a result in disharmony (well-knowing that some topics are diametric and real consensus cannot be created)? How do we judge a communicative planning process? On a very low level this is possible, e.g. if there has been a debate about the topic at all or if different demands have been considered. But to decide whether planning has been effective or not gets even more complicated because planning authorities are not the implementing bodies (Einig, 2012; Weith, 2007).

- This is possibly reflected in the urban planning authority's self-conception as well. As the authority is not acting like one organization with one organizational culture and one obvious target to implement, the self-conceptions vary significantly (as they also do in the two case studies). On the one hand, urban planning authorities can be crucially influenced by political statements, positioning and strategies (as in the example of Nuremberg), and, on the other hand, the collaboration between different departments of the same administrative bodies has to be checked out and an informal, inter-departmental integration is a precondition of further planning. But planning is normative; planning implies the handling of societal values and norms about how the urban society wants to live in future. There is no 'right' or 'wrong', there is a kind of debate, consensus or not, and many other demands that have to be respected and considered. Additionally, urban planning authorities are seen as advocates for the common good and public welfare. So, all in all, the urban planning authorities are characterized by a hybrid situation between high claims of coordination, the evaluation of different demands and speaking for the common good, in a cultural context of more than one organization (department), paired with weak instruments and no possibilities for implementation. Their self-confidence depends very much on the organizational leadership, the functions and tasks they occupy as well as their freedom of decision-making and their willingness to be involved.

Shown by the examples of Ratisbon and Nuremberg, the different organizational cultures of both land and water governance were sketched. These different cultural imprints of the two main actors determine to a high degree the established planning culture.

The added value of a cultural perspective is, first of all, to state that culture matters, even in planning processes where legal bases, standardized instruments and European-wide assessments play an important role. However, sometimes it is even the cultural embeddedness of management processes that makes them work, or, as Peter Drucker is supposed to have claimed, 'Culture eats strategy for breakfast and technology for lunch [...]' (as cited in Fields, 2006). Second, the possibility that considering the cultural perspectives of planning can lead to a further understanding of planning processes, decision-making and cooperation. Called the 'black box of integration', it is still not empirically analysed how planners integrate different aspects and how they can come to a deeper understanding of the cooperation-partner logic. The cultural analysis of planning processes is based on individual to organizational phenomena, on all kinds of levels with special regard to their cultural implementation and imprint. Methodologically entering new ground, the analysis of cultural aspects of planning processes is largely unexplored, as culture is omnipresent, not able to be isolated, and becomes evident in nearly all aspects, which discourages many researchers. On the other hand, the possibility for a deeper understanding of planning processes is there. As a final note, it should be mentioned that, of course, considering a cultural perspective does not lead to the possibility of steering planning processes though culture. As Schein (2003) has outlined already, culture is not a means of steering: only the consequences of culture can be handled and managed. This is an important fact, as culture and its consequences have to be understood and comprehended before one is in a position to manage it.

Notes

1. IWRM has been defined by the Global Water Partnership (2000) as 'a process which promotes the coordinated development and management of water, land and related resources, in order to maximize the resultant economic and social welfare in an equitable manner without compromising the sustainability of vital ecosystems' (p. 22). This approach is often used by water management research projects, and sometimes by environmental science researchers, but less frequently in spatial planning. In the context of this article, IWRM is associated with an integrative claim of riverscape development, but in a much broader field (as water resources include much more than just rivers).

2. Referring to the (critical) discussions about a coastal manager (Billé, 2008) or a landscape manager (Pryce, 1991), a river manager is not per se desirable, either. The strong point of interest behind such an institution is to foster the integration process in a formal way instead of accepting a multilayered system typical for its complex allocation of roles.

References

Benz, A., Lütz, S., Schimank, U., & Simonis, G. (2007). *Handbuch Governance: Theoretische Grundlagen und empirische Anwendungsfelder*. Wiesbaden, Germany: Verlag für Sozialwissenschaften.

Billé, R. (2008). Integrated coastal zone management: Four entrenched illusions. *S.A.P.I. EN.S, 1*(2), 75–86.

Biswas, A. K. (2004). Integrated water resources management: A reassessment. *Water International, 29*(2), 248–256. doi:10.1080/02508060408691775

DiGaetano, A., & Strom, E. (2003). Comparative urban governance: An integrated approach. *Urban Affairs Review, 38*(3), 356–395. doi:10.1177/1078087402238806

Driessen, P. P. J., Dieperink, C., van Laerhoven, F., Runhaar, H. A. C., & Vermeulen, W. J. V. (2012). Towards a conceptual framework for the study of shifts in modes of environmental governance – experiences from The Netherlands. *Environmental Policy and Governance, 22*(3), 143–160. doi:10.1002/eet.1580

Einig, K. (2012), Evaluation in der Regionalplanung. *Informationen zur Raumentwicklung, 1/2.2012*, I–IV.

Ernste, H. (2012). Framing cultures of spatial planning. *Planning Practice & Research, 27*(1), 87–101. doi:10.1080/02697459.2012.661194

Faust, T. (2002). *Organisationskultur und Ethik Perspektiven für öffentliche Verwaltungen*. Berlin, Germany: TENEA.

Friedmann, J. (2005). Planning cultures in transition. In B. Sanyal (Ed.), *Comparative planning cultures* (pp. 29–44). New York, NY: Routledge.

Fürst, D. (2001). Regional governance – ein neues Paradigma der Regionalwissenschaften? *Raumforschung und Raumordnung, 59*(5–6), 370–380. doi:10.1007/BF03183038

Fürst, D. (2007). Planungskultur: Auf dem Weg zu einem besseren Verständnis von Planungsprozessen? PNDonline, III/2007. Retrieved from http://www.planung-neu-denken.de/ images/stories/pnd/dokumente/pndonline3-2007-fuerst.pdf

Gleick, P. H. (2000). A look at twenty-first century water resources development. *Water International, 25*(1), 127–138. doi:10.1080/02508060008686804

Global Water Partnership. (2000). *TAC background Papers No.4: Integrated water resources management.* Stockholm, Sweden: Global Water Partnership.

Gullestrup, H. (2009). Theoretical Reflections on Common European (Planning-)Cultures. In J. Knieling & F. Othengrafen (Eds.), *Planning cultures in Europe: Decoding cultural Phenomena in urban and regional planning.* London, UK: Ashgate Publishing.

Hartmann, T. (2010). Reframing polyrational floodplains: Land policy for large areas for temporary emergency retention. *Nature and Culture, 5*(1), 15–30. doi:10.3167/nc.2010.050102

Healey, P. (1996). The communicative turn in planning theory and its implications for spatial strategy formations. *Environment and Planning B: Planning and Design, 23*(2), 217–234. doi:10.1068/b230217

Hutter, G. (2007). Strategic planning for long-term flood risk management: Some suggestions for learning how to make strategy at regional and local level. *International Planning Studies, 12*(3), 273–289. doi:10.1080/13563470701640168

Knieling, J., & Othengrafen, F. (2009). En route to a theoretical model for comparative research on planning cultures. In J. Knieling & F. Othengrafen (Eds.), *Planning cultures in Europe: Decoding cultural phenomena in urban and regional planning.* London, UK: Ashgate Publishing.

Levin-Keitel, M., & Sondermann, M. (2014), Planerische Instrumente in lokalen Kontexten – Einblicke in die Vielfalt von Planungskulturen. *Arbeitsbericht der ARL, 10*, 172–191.

Levin-Keitel, M. (2014). Flusslandschaften in der Stadt. *RaumPlanung, 172*(1–2014), 20–25.

Mitchell, B. (2005). Integrated water resource management, institutional arrangements, and land-use planning. *Environment and Planning A, 37*(8), 1335–1352. doi:10.1068/a37224

Molle, F. (2008). Nirvana concepts, narratives and policy models: Insights from the water sector. *Water Alternatives, 1*(1), 131–156.

Moss, T. (2009), Zwischen Ökologisierung des Gewässerschutzes und Kommerzialisierung der Wasserwirtschaft: Neue Handlungsanforderungen an Raumplanung und Regionalpolitik. *Raumforschung und Raumordnung, 67*, 54–68. doi:10.1007/BF03183143

Nuissl, H., & Heinrichs, D. (2011). Fresh wind or hot air – does the governance discourse have some-thing to offer to spatial planning? *Journal of Planning Education and Research, 31*(1), 47–59. doi:10.1177/0739456X10392354

Nuissl, H. (2008). Umfrage zur 'Planungskultur'. *PNDonline, I/2008.* Retrieved from http://www. planung-neu-denken.de/images/stories/pnd/dokumente/pndonline%204%202007_umfrage.pdf

Othengrafen, F. (2012). *Uncovering the unconscious dimensions of planning. Using culture as a tool to analyse spatial planning practices.* Farnham, UK: Ashgate Publishing.

Pryce, S. (1991). Community control of landscape management. *Planning Outlook, 34*(2), 75–82. doi:10.1080/00320719108711896

Reimer, M., & Blotevogel, H. H. (2012). Comparing spatial planning practice in Europe: A plea for cultural sensitization. *Planning Practice & Research, 27*(1), 7–24. doi:10.1080/ 02697459.2012.659517

Rittel, H. W. J., & Webber, M. M. (1973). Dilemmas in a general theory of planning. *Policy Sciences, 4*(2), 155–169. doi:10.1007/BF01405730

Scharpf, F. W. (2000). *Interaktionsformen: Akteurzentrierter Institutionalismus in der Politikforschung.* Opladen, Germany: Leske + Budrich.

Schein, E. H. (2003). *Organisationskultur: The Ed Schein Corporate Culture Survival Guide.* Bergisch Gladbach, Germany: EHP.

Stadt Nürnberg. (2012a). Nürnberg in Zahlen 2012. Amt für Stadtforschung und Statistik für Nürnberg und Fürth, Nürnberg.

Stadt Nürnberg. (2012b). Integriertes Stadtentwicklungskonzept Nürnberg am Wasser. Wirtschaftsreferat, Amt für Wohnen und Stadtentwicklung, Nürnberg.

Stadt Regensburg. (2012). Regensburg in Zahlen. Ausgabe 2012. Amt für Stadtentwicklung, Abteilung Statistik, Regensburg.

United Nations Educational, Scientific and Cultural Organisation. (2014). World Heritage List. The old town of Regensburg with Stadtamhof. Retrieved from http://www.whc.unesco.org/en/list/1155

Von Haaren, C., & Moss, T. (2011). Voraussetzungen für ein integriertes Management: Koordination und Kooperation der wasserrelevanten Akteure und Organisationen in Deutschland. In C. Von Haaren, & C. Galler, (Eds.), *Zukunftsfähiger Umgang mit Wasser im Raum* (pp. 67–81). Hannover, Germany: Akademie für Raumforschung und Landesplanung.

Wasserwirtschaftsamt Regensburg (WWA) und Stadt Regensburg. (2010). Hochwasserschutz Stadt Regensburg. Flussraumkonzept Donau – Regen. Regensburg.

Weith, T. (2007). Anforderungen an Evaluationen von Umbauaktivitäten in Städten und Regionen. In T. Weith, (Ed.), *Stadtumbau erfolgreich evaluieren* (pp. 237–252). Münster, Germany: Waxmann Verlag.

Wiering, M., & Immink, I. (2006). When water management meets spatial planning: A policy-arrangements perspective. *Environment and Planning C: Government and Policy, 24*(3), 423–438. doi:10.1068/c0417j

Legal frameworks

Baugesetzbuch (BauGB) in der Fassung vom 23.09.2004 (BGBl. I S. 2414) zuletzt geändert durch Gesetz vom 11.06.2013 (BGBl. I 2005, 1548).

Directive 2007/60/EC of the European Parliament and of the Council of 23 October 2007 on the assessment and management of flood risks (Flood Directive).

Gesetz zur Ordnung des Wasserhaushalts 2010 (Wasserhaushaltsgesetz) (WHG) vom 31. Juli 2009 (BGBl. I 2009, 2585).

The governance dilemma in the Flanders coastal region between integrated water managers and spatial planners

Karel Van den Berghe and Renaat De Sutter

Centre for Mobility and Spatial Planning, Ghent University, Ghent, Belgium

The Flemish coastal region has two major key challenges: coastal flood risk and risk of drought. As an answer to the first challenge, a new phase of land reclamation on sea is proposed, fitting into its historical path dependence. This, however, will aggravate the second challenge, and contradict the principles of integrated water resources management (IWRM). The two challenges take place on two different governance frontiers of land and water governance, but have a growing mutual influence. It is argued that the coastal spatial governance regarding these two challenges suffers from a lock-in.

Introduction

The coastal area in Flanders is an approximately 10–20 km-wide reclaimed lowland, at an elevation that varies from 2.5 to 5.5 m LAT.[1] The 65 km of shoreline is bordered by an almost continuous coastal dune belt and crossed by a small river, the Yser, which is presently canalized and flows into the North Sea at the town of Newport (Baeteman, Scott, & Van Strydonck, 2002; Ervynck et al., 1999). The plain is subdivided into three parts: low-lying polder areas, dunes and shore, with the low-lying polder areas forming the major part of the coastal plain (Lebbe, Van Meir, & Viaene, 2010). The current coastal plain can be identified as a flat, open polder area consisting mainly of intensive agricultural landscapes and widespread small villages (Antrop, 2007). Only along its coast on a stretch of land of approximately 0.5–2 km wide, situated mostly on the belt of coastal dunes and dikes, is there a high densification of tourism, mainly consisting of high and dense apartment buildings and villas (Figure 1) (Antrop et al., 2002; Koks, de Moel, & Bouwer, 2012). In total along the coastline, only 40% is naturally protected by the dunes. Elsewhere, additional soft or hard artificial defences are needed (Lebbe et al., 2010). Without these dikes and dunes along the coastline, most parts of the coastal area would float every day due to its low-lying profile (EEA, 2006).

Looking at the water configuration of the Flemish coastal region, it is remarkable that, besides the small river Yser, there are no large rivers flowing from the inland into the North Sea. The configuration of the current river system has its origins around 7 million years ago, when the sea was situated to the north of Belgium with a west–east coastline. Since then, except for the Yser, no river has changed it course towards the more recently developed North Sea (Broothaers, 1995). This specific geological evolution has led to the situation that, with exception of local precipitation, there is no natural major net flow of

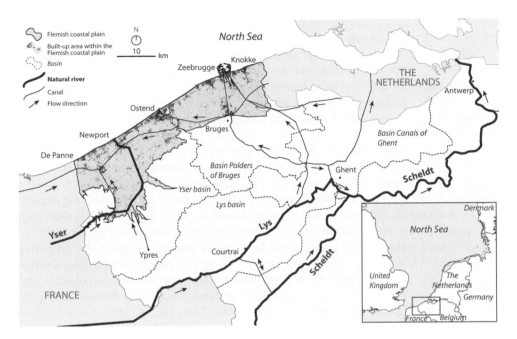

Figure 1. The Flemish coastal plain.

fresh water from inland to the coastline. Moreover, when there is a dry period, the net discharge of the Yser falls to zero or is even negative. During a dry period, there is not enough fresh water to satisfy current water demands. Even the ground water reserves are limited (VMM, 2006). In these periods it is vital that there are transfers of fresh water through canals towards the coastal region. This is a complete different situation compared with, for example, the River Rhine in the Netherlands. Even in dry periods, due to its huge river catchment, the Rhine carries enough water to satisfy the water demands. The average discharge volumes of the Yser, Lys and Rhine are around 8, 40 and 2300 m^3/s respectively (Bursens & Puype, 2011; Heylen, 1997; PBL, 2011).

The Flemish Decree of Integrated Water Policy states that the impact of the water cycle on the water system should be minimized and human interference only tolerated (CIW, 2003). This decree is the implementation of the European Water Framework Directive (EC, 2000). According to Gleick (2000) these new directives are part of the changing water paradigm. The traditional planning approaches relied on physical solutions. However, these measures are facing increasing problems, especially in the context of the present climate change. In the case of the Flemish coastal area, the climate change will, next to the rising sea level and the related higher risk of coastal flood, lead to a change in seasonal precipitation patterns (IPCC, 2014). A possible decrease in summer precipitation combined with an increase in temperature would lead to loss of availability of water during the summer (Brouwers et al., 2009; De Sutter, 2011). These dry periods will cause the need of more transfers of fresh water to the coastal region in future (Allaert, 2012). Therefore the need for human interference in the water cycle will only increase, thus going even more against the dogma of integrated water resources management (IWRM).

When analysing the current water system of the Flemish coastal plain, one might wonder how it is possible that this region can exist with such great human interference. This shows that when analysing the current coastal area, there is a need seriously to take into account its past. Today the big question is how the coastal area should evolve during the 21st century. On the one hand, there is need for protection from coastal floods, and, on the other hand, there is the need of additional access to fresh water for the coastal area (EEA, 2006). To deal with these two situations, only technical solutions and thus more human interference is the solution, meanwhile moving away from self-sustainability. It is clear that policy-makers have to find a sustainable solution for the dilemma between those two opposite principles – the hold-the-line principle and the IWRM principle – to answer future problems.

To handle this dilemma, the first section uses a profound historical review based on new insights to sketch the socio-economic evolution of the Flemish coastal region. Hereby we try to construct as well as possible the path dependence of this evolution. Following this section, we describe the current situation and future challenges regarding the climate, the socio-economic and policy change. In the discussion and conclusion we indicate that the current policy developments are just a new episode in the path dependence of the coastal spatial policy. The current coastal boundary is assured. However, by proposing this, the other less notable boundary of summer drought is pushed more inland. Current spatial policy does not provide a good balance between the socio-economic and physical–ecological aspect of the coastal region. An important role in this multidisciplinary discussion could be taken up by spatial planners. Therefore, we state that the coastal spatial policy is suffering from a lock-in.

A historical review on the constantly changing spatial policy of the coastal region

Throughout its history, the spatial configuration of the Belgian coastal region has changed fundamentally. Its history and its water policy are a perfect example of path dependence (Crabbé, 2008): actions are building further on decisions made previously. The integration of water management history and economic history is a relative recent but important turn in historiography (Soens, 2006). This section makes an objective analysis based on new insights to explain how the current coastal plain and its hydrological cycle have been formed.

Living in an unliveable coastal area

The first signs of human presence in the Belgian coastal region date from the Iron age (2700 BP). Historians now are convinced that since the Roman period there has always been a permanent occupation of the Belgian coastal region (Loveluck & Tys, 2006). Moreover, due to the weakening of the inland authority by raids of the Norsemen in the 9th and 10th centuries, there was an establishment of successful North Sea people who became rich by trading between all the coastal regions around the North Sea (Crabtree, 2010; Tys, 2005).

The take-over of the central authority and the first (economical) land reclamations

The coastal environment thus incited the coastal society to become rich by trading the unique natural resources for missing resources, like wood. Also the breeding of sheep on the biologically rich marshes and the processing of cloth was a very lucrative business. The big question therefore is why the decision was made to reclaim land from the sea and to change the natural richness which at first made permanent occupation possible.

This change in policy seems to be a result of a combination of events. The central authority prevented the Norsemen from conquering the land by establishing fortifications around the major rivers and on several positions in the coastal area. By this, it reclaimed its power over the coastal area. Nevertheless, the Count of Flanders became really powerful due to a natural event. In the 11th century, parts of the estuarine area of the Yser were silting up (Verhulst, 2000). This natural event gave birth to new marshes. These new marshes were economically very interesting for sheep breeding because of the popularity of Flemish cloth. By law, all this new land became property of the Count of Flanders. Through leasing and selling these marshes the count became a very powerful authority in Western Europe (Declercq, 2000).

In the 12th century the rate of natural emergence of land was, however, too low to satisfy economic needs. The Count of Flanders gave permission for big institutions, such as abbeys, to start building offensive dikes and to reclaim land from the sea (Crabbé, 2008). These first embankments were thus a consequence of economic needs. Protected by dikes, land that normally would periodically flood came available for agriculture year in year out (Declercq, 2000).

At first, the embankment was organized on a local base, every inhabitant helping to maintain his own polder with dikes. As the embankments became bigger, polders started to work together. Therefore, from the 13th century onwards, water boards were established. These water boards took over the maintenance of the dikes (Crabbé, 2008).

Large-scale landownership contributed to numerous inundations

The evolution towards water boards is an important fact because it marks an evolution in social stratification. Due to the water boards, a water elite class emerged where democratic values quickly disappeared. There was inheritance of the water board seats, so families became mightier and mightier through leasing and buying of new lands (Soens, 2009).

Due to a worsening economic situation at the end of the 13th century, increasingly farms could not afford the rising costs and had to sell land. Therefore, many farms became the property of a few owners, mostly non-farmers. Both institutional and bourgeois landownership were thus gaining importance. As a consequence, most of the landowners in the 16th century did not live in coastal areas, but in cities. The biggest concern for these investing bodies was making money through leasing of their property (Soens, 2011). Tenants were at the mercy of the purely economic-thinking landowners. If the costs were greater than the profits of a polder, they simply stopped their investments in it. The landowners gave up their least profitable properties (Soens, 2006). The dikes breached and the polders became wetlands again until there was enough money to reinvest in its dikes (Thoen & Soens, 2003).

The commercial network of canals made transfer of water possible

The economic success of Flemish cloth led to the rising of the powerful medieval cities of Bruges, Ghent and Ypres, among others. To trade with each other and with other European regions, they needed a constant and navigable access towards each other and towards the North Sea. Their only concern was trade and transport (Crabbé, 2008). At first it was sufficient to connect the different creeks with each other, but due to the enlarging of the ships, soon it was necessary to dig canals. In a few centuries, an enormous network of canals was installed in the Flemish coastal plain (Figure 1).

From this point of there were two institutions with different goals responsible for the water configuration in the coastal plain. The water boards, on the one hand, attempted to keep their polders dry and tried to drain the water flows as fast as possible to the sea to promote farming and to prevent inundations. The cities, on the other hand, wanted to keep the canals at all times navigable by keeping a minimum water level in the canals (Soens, 2006, 2009).

Next to the commercial aspect of the canal network, these new connections made water transfers over long distances possible. The most important and longest canal is the Ghent–Bruges–Ostend–Newport canal. It is approximately 70 km long and connects the river basin of the Lys with the coastal region (Figure 1). This connection was and still is essential. Due to the lack of big natural water inflows towards the coastal region and the need of a constant water level in the canals, there is also today a constant transfer of water from the Lys through this major canal towards the other canals in the network of the coastal region (Van den Berghe & De Sutter, 2014).

The rising tourism sector and the drinking water network

As a trend that started in England, coastal tourism appeared first in Ostend (Charlier & De Meyer, 1992). Boosted by King Leopold II, who built his Royal Villa in Ostend in 1873, there was, until the beginning of the First World War in 1914, a continuous construction of hotels along the coastline. At the same time other coastal villages, as, for example, Knokke, influenced by the elite of Bruges, developed as seaside resorts (Gobyn, 1987). Later on, planners designed completely new townships in different dune areas. To protect this new heritage against the sea, dikes were built and associated promenades were constructed (IMCORE, 2009).

The growth of tourism resumed after the Second World War. During the period 1960–90s, tourism increased by 38%. As a consequence, the expansion of coastal towns towards each other has resulted today in ribbon development along almost the entire coastline (Figure 1) (Koks et al., 2012; Ruyck, Ampe, & Langohr, 2001).

The tourism sector along the coastline had an important influence on the drinking water configuration of the coastal plain. The only useable locations at the coast where drinking water can be extracted are the rather small freshwater lenses in the coastal dune massifs at the western and eastern coasts. The rest of the coastal area has almost no useable groundwater reserves. For centuries, the only way to have decent drinking water in more dense places was by collecting rain water (Valcke, 1992). However, the many rich tourists in the beginning of the 20th century, mostly elite from the capital city of Brussels, wanted to have the same quality of drinking water as at home and, moreover, they wanted a drinking water network under pressure. Therefore, in 1932, delayed by the First World War, the construction of the Pipeline of Flanders between Brussels and Ostend was finished. Through this pipeline, the cities of Bruges and Ostend, among others, received drinking water from the region of Brussels (Valcke, 1993, 1994).

As tourism grew, more drinking water needed to be distributed. Therefore today, especially during the summer period, when in general the availability of water is scarce and coastal tourism water demand is high, enormous transfers of drinking water occur toward the coastal area (SVW, 2008; Van den Berghe & De Sutter, 2014).

The most recent inundation

In general coastal flood hazards are considered worldwide as one of the most significant natural disasters in terms of human impact and economic losses (Jonkman & Vrijling,

2008). In Belgium, the most recent severe coastal flood occurred in 1953. This storm struck the coastal areas of the North Sea. Many deaths and enormous damage to the infrastructure was the consequence (Gerritsen, 2005). After this event, an important part of the Belgian sea walls was reinforced (Charlier & De Meyer, 1992). Policy clearly chose the hold-the-line scenario. As a result of the big defence investments, since 1953 no coastal flood has occurred. This had two consequences. First, an awareness of an extreme and unsafe situation is fading (Colten & Sumpter, 2009). Especially, younger people have a false sense of safety from coastal floods (Kellens, Zaalberg, Neutens, Vanneuville, & De Maeyer, 2011). Secondly, because of this sense of safety, investments in property along the coastline have increased, making the tourism and economic sector increasingly important.

The current situation and future challenges

The economic structure of the coastal area

In Flanders, as everywhere in Europe, population densities along coastlines are higher than those elsewhere in the coastal area (EEA, 2006). At peak moments there is a population density of no fewer than 2830 persons/km^2 (Monballyu, Van den Driessche, & Pirlet, 2014). The built-up area in the 67 km long and 0–1 km wide coastal strip is approximately 50%, much higher than, for example, France (25%) or the Netherlands (12%) (EEA, 2006). The Flemish coastal area can therefore be indicated as a long, stretched urban area; this in high contrast to the immediate open area just a few kilometres further inland (Coudenys et al., 2014). The coastal area occupies a major part in the total Flemish touristic circuit. A total of 35.6% of the total overnight stays in 2008 in Flanders were at the coast (Bourgeois, 2009). In total, the coastal tourism sector had in 2012 a direct added value of €2.9 billion (Westtoer, 2013).

Along the coastline there are two economic ports, Ostend and Zeebrugge. Of the total direct added value in 2011 (€16.482 million), 5.8% was realized by the port of Zeebrugge (€956 million) and 2.9% by Ostend (€478 million). These two ports are rather small. Most of the industry in the coastal area is linked to the port functions as part of an evolution ongoing in the whole of Europe (EEA, 2006).

Besides the urbanized coastal strip, the coastal area is mainly an agricultural area (70.761 ha). Historically, as in the rest of Europe, this activity has been very important (EEA, 2006). As in all European member states, the share in gross domestic product and in employment of Flemish agriculture is very low and declines year after year (Van Passel, Nevens, Van Huylenbroeck, & Mathijs, 2005). In 2012 the share in gross domestic product in Belgium was no more than 0.75% (Belgian Federal Government, 2013). This is slightly higher in the coastal region, but does not exceed 3% (WES, 2009). Since 1962 the agriculture economy is strongly subsidized. In 2007 the total subsidy per hectare in Flanders was around €500, which is significantly higher than the average of about €300/ha in EU-27 countries (Belgian Federal Government, 2011).

In summary, the major key player in the economic structure in the coastal region is, when considering relevant indicators as direct added value and employment, the tourism sector followed by the economic ports of Ostend and Zeebrugge. Spatially, the two economies do not take much space in the whole coastal area. Still, the biggest part is, as it was for centuries, mostly the agricultural area. However, seen from an economic point of view, this sector is not an example of a sustainable sector due to its subsidy support.

Water transfers

Currently there are two major water transfers in the coastal area. The first transfer is surface water via the canal network. To understand fully how this water network is managed, it is necessary to take into account three possible situations: a normal situation, an extreme dry situation and an extreme wet situation (Van den Berghe & De Sutter, 2014). On an average base, there is about 1750 m^3/year/capita fresh water available in Belgium (FAO, 2003). However, in Flanders, moreover in the Scheldt catchment, this is about 650 m^3/year/capita much less (VMM & CIW, 2009). Per definition, a region is scarce of water if the annual average (theoretically relying on its internal resources) is below 1000 m^3/capita (FAO, 2003). Keeping into account that in a dry situation the net flow of fresh water drops and that there are no considerable ground water reserves, the coastal region is therefore a very scarce water region. In this situation the coastal region is vitally dependent of the water inflow from the Lys, the closest natural river. To maintain a sufficient water level, surface water from the Lys is transported over a distance of almost 100 km through the Ghent–Bruges–Ostend canal by which it supplies the other canals and even the Yser with fresh water (Van den Berghe & De Sutter, 2014).

The second major water transfer is the transfer of drinking water towards the coastal region. This transfer is impressive. Taking into account that a person's daily demand for water is constant at about 100 litres/day, it is clear that there is a need for huge transfers of drinking water, especially during the summer period. On an annual basis, no fewer than about 20 million m^3 of fresh water are transferred to the coastal region of Flanders (Van den Berghe & De Sutter, 2014).

The rise of regulation in water policy

As a concept, IWRM has become a powerful and all-embracing slogan over the past 20 years (Biswas, 2004). A widely accepted definition of IWRM is that of the Global Water Partnership (GWP, 2014). This definition describes IWRM as 'a process which promotes the coordinated development and management of water, land and related resources in order to maximize the resultant economic and social welfare in an equitable manner without compromising the sustainability of vital ecosystems' (Cook & Spray, 2012; GWP, 2000; Mitchell, 2005). IWRM is used to emphasize the integration and balance of social and physical factors and is used to explore complex human–environment relations in order to implement a successful policy (Cook & Spray, 2012).

In Europe, the first wave of regulations about water started in 1975 with the European Water Legislation with standards for rivers and lakes used for drinking water abstraction (EC, 1975), followed by the Urban Waste Water Directive and The Nitrates Directive in 1991 (EC, 1991). As the Drinking Water Directive and the Urban Waste Water Directive were considered as milestones, in 1996 the European Parliament requested the Commission to come forward with a proposal for a Water Framework Directive. This resulted in 2000 in the Water Framework Directive (EC, 2000; Rahaman, Varis, & Kajander, 2004). The river basin approach stood centre stage. With this directive, all member states had to define and implement the necessary measures within their integrated programmes (EC, 2000).

Having a regional competence, Flanders was also required to translate the Water Framework Directive into a decree. This decree was approved in 2003 (CIW, 2003). According to the European Directive, the Flemish Decree is based on the river basin approach. This was, however, difficult in areas where the water flow is technically organized. Especially in the coastal area, the division into river basins was difficult and per definition artificial (Crabbé, 2008). Eventually the coastal area was divided into two large river basins: the basin of the Yser in the south-west and the basin of the polders of Bruges in the north (Figure 1) (VMM, 2008a, 2008b).

Future challenges

The climate scenarios

There are two major key challenges for the coastal area due to climate change. Sea level rise is the most obvious. According to the IPCC (2014), the increasing coastal flood risk will remain a key challenge. Without any adaptation, the risk of coastal flooding in the 2080s will increase, especially for the Atlantic, Northern and Southern European regions. Countries with high absolute damage costs are, among others, the Netherlands, Germany and also Belgium. Indeed, sea-level-rise monitoring shows that the high water level has risen about an average of 20 mm/decade and mean sea level an average of 15 mm/decade over the 20th century (De Sutter, 2011; Lebbe et al., 2010). Hence, the high water level is increasing faster than the low water level. This indicates an increase in tidal amplitude (Lebbe et al., 2010), meaning that the impact of waves during a storm will be significantly greater than now (Van den Eynde, De Sutter, & Haerens, 2012).

Next to sea level rise is the, not always known, constantly increasing problem of drought due to changing precipitation patterns. As noted above, the drought problem is already a problem for the Flemish coastal area. Moreover, even without a further increase in the variability of precipitation patterns, due to the socio-economic change, the demand of fresh water will increase (Allaert, 2012). Therefore, according to De Sutter (2011), this could be the key challenge for the future. For example, 2009 has been the driest year in Belgium since records began. Studies about the projection of drought expectations are not available at the level of the Flemish coastal plain, but it is likely to be expected that in combination with the increase in heatwaves, the imbalance between demand and supply of fresh water will become greater (De Sutter, 2011).

Socio-economic scenarios

Future vulnerability cannot be predicted only based on climate scenarios. Changes in the socio-economic situation are at least as important. In the CcASPAR project[2] four socio-economic scenarios were used to show the future land-use projection for the coastal area of Flanders (De Moel et al., 2012; IPCC, 2000; Koks et al., 2012). In all future scenarios there is a stronger pressure on the available land. This pressure is the result of, depending on the scenario, mainly extra residential uses (A1), extra agriculture (A2), extra urban areas (B1) or an extra need to preserve natural areas (B2) (De Moel et al., 2012). It is estimated that when the sea breaks through the coastline, potentially 200,000 people could be affected (De Sutter, 2011).

Based on the climate and socio-economic scenarios, the Flemish coastal region is a winner of climate change. Weather conditions will improve and probably trigger a longer and more crowded tourism season (Nicholls & Amelung, 2008). Investing in tourism infrastructure could be lucrative in the long-term. These investments will lead to two major problems. First, the investments as a result of better touristic conditions, but also the presence and expansion of the ports of Ostend and especially Zeebrugge will increase both the economic and social values along the coastline. To minimize the risk of major damage due to coastal floods, more investments in coastal defences will be necessary. Second, the demand for fresh water for a range of activities will rise. In combination with the decreasing constant presence of fresh water, this will lead to more and larger transfers of both drinking water and surface water through, respectively, pipes and canals from the inland towards the coastal region. The water system at the coastal area, which is mainly human made, will need more human interference, deviating even more from a self-sustainable situation.

Future policy

Recently, two new government documents set out the spatial coastal and maritime policy (Figure 2). The Belgian Federal Government has the authority over the Belgian part of the North Sea. Notwithstanding the rather small area (3454 km^2; 0.5% of the North Sea), this part of the North Sea is one of the busiest in the world. Activities are numerous and varied. To fit all these activities together, the federal authority drew up a marine spatial plan for the Belgian part of the North Sea (Figure 2b) (Belgian Federal Government, 2014). The style of this spatial plan is in between a vision plan and a structure plan.

The Flemish Government has the authority over the coastal area and in 2011 approved the Master Plan for Flanders Future Coastal Safety, with a total investment of €300 million. The purpose is to reach an adequate protection level for a so-called millennial storm. The measures are planned between 2011 and 2015 and consist of soft measures, like beach nourishment, and hard measures, like storm return walls (Mertens et al., 2011). Next to this 2050 plan, a long-term vision with a time horizon of 2100 was developed in 2012 called the Flemish Bays (Flemish Government, 2012). The central principle is a sustainable coast growing with the sea. Notable is that to reach this goal, the plan must, first, create new land offshore and, second, obtain and improve accessibility and ensure the growth of the ports of Ostend and Zeebrugge. Recently, the plan has been translated into a more specified plan (Figure 2a) (THV Vlaamse Baaien, 2014). The most striking element in the plan is the creation of islands. These islands are seen as a solution (1) to

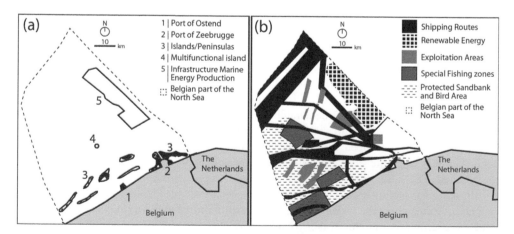

Figure 2. Belgian part of the North Sea: (a) The Flemish Bays and (b) The Marine Spatial Plan of the North Sea. Source: based on: Belgian Federal Government (2014) and THV Vlaamse Baaien (2014).

give more space for the different activities along the coastline, (2) to act as breakwater for incoming waves and (2) to store wind energy. Their unique selling point is that investing in coastal safety is combined with economic profit.

Discussion and conclusion: the utopia of IWRM and the lock-in of economic driven spatial planning

Currently the spatial water issues are regulated by different river basins. As Gleick (2000) states, this is part of a changing water paradigm that has been going on since the millennium change. The challenge is to reintegrate water use by maintaining ecological health and environmental well-being. However, in this specific case we think this is a utopia. The demand of water is just too high and the transfer of surface and drinking water from inland is and will always be necessary. The only solution to raise the supply of water locally is to build more reservoirs or more pipelines, which is the technological solution.

The coastal area originated from agricultural economic desires and this analysis shows that this still is the main motive of spatial policy, although the economic focus has now shifted to tourism and port development. In spatial terms, this new focus area is a rather small area boarding the coastline. The residual agricultural part of the coastal area is in economic terms much less valuable. According to Tempels & Hartmann (2014, this issue), this is a good example of the dike paradox: value is accumulated behind, or, in this case literally on, dikes as a consequence of the flood protection measures.

To project this high economic coastline, the hold-the-line principle is quite logical. However, due to climate change this principle is not sufficient anymore. The most recent policy is therefore to expand the line and develop new formed islands. This plan fits perfectly in the path dependence of its spatial planning history and will, if executed, announce a new episode in reclaiming land from the sea.

The Flemish Bays will have to tackle two problems. The first is a typical land-use conflict that will arise in the North Sea, if one relates this plan to the Marine Spatial Plan. This plan focuses mainly on the sea itself and does not take into account coastal defence. The line of islands proposed by the Flemish Bays could interfere with the activity areas in the Marine Spatial Plan. These spatial conflicts are perfect examples of economic-driven spatial planning; each institution, in this case the Flemish and Belgian federal governments, wants to secure the available space for its own main purposes. Who profits from the Flemish Bays (Hartmann & Tejo, 2014, this issue)? The starting point of the plan is climate change challenge. This analysis, however, shows that this might be considered by some stakeholders as a covering story and enlarging the economic patrimony is their major goal. The second less notable problem is the increasing drought problem pushing the inland boundary of drought more inland.

There are two horizontal boundaries. The first is the most obvious: the coastline where the land meets the sea. The second is dependent on a balance between the demand and supply of fresh drinking and surface water, is less static and is situated more inland. As Hartmann & Tejo (2014) state, along these boundaries governance frontiers occur. Especially along the coastline, a conflict will probably appear in future.

Seen from a socio-economic point, the current situation is quite comparable with what happened in earlier centuries. Today, the agricultural polders have no sustainable economic value, of course making abstraction of its landscape or historic value among others. Different now is that the maintenance of the dikes is not stopping. Today, the dikes and the coastal strip have a larger economic value than the coastal plain protected by the dikes. Hypothetically, if the economic shift would not have occurred, the maintenance of the dikes would have stopped and the polder area would have become tidal area again, as has happened a few times in history.

One may think how spatial planning can deal with this dilemma between the economic situation and the non-sustainable water system. This is of course not easy. Certain is that fresh water availability will decrease. The proposed coastal defence strategy, however, may be revised to meet the IWRM principles. It is quite remarkable when seen from a purely economic point of view that to protect the coastline the less economically valuable hinterland has also to be protected. In the research project CcASPAR this was worked out (Allaert, 2012, pp. 166–187). The starting point was not to take the coastline, and hence hinterland, as one entity, but as several compartments. This strategy offers a more flexible framework where more climate impacts can be integrated. This way of thinking is not investigated in current policy documents. Therefore, we state that the coastal spatial planning is suffering from a lock-in due to its spatial path dependence. Hold the line is the only option and the problems of coastal flood and drought are not combined. To deal with this multidisciplinary problem, spatial planners can play an important role in composing a vision where land and water can harmonize in a more resilient way. However, we presume that this will not change soon and only a new tipping point, e.g. a huge storm or long drought period, could bring a changing paradigm in coastal spatial management policy.

Notes

1. Lowest astronomical tidal prediction in a time span of at least 18.6 years (MDK, 2014).
2. Climate change and changes in spatial structures research project (http://www.ccaspar.ugent.be/).

References

Allaert, G. (2012). *Klimaat in Vlaanderen als ruimtelijke uitdaging (CcASPAR)*. Ghent: Academia Press.

Antrop, M. (2007). *Perspectieven op het landschap: achtergronden om landschappen te lezen en te begrijpen*. Ghent: Academia Press.

Antrop, M. et al. (2002). *Traditionele Landschappen Vlaanderen: Kenmerken en beleidswenselijkheden*. Belgium: Department of Geography, Ghent University.

Baeteman, C., Scott, D. B., & Van Strydonck, M. (2002). Changes in coastal zone processes at a high sea-level stand: A late Holocene example from Belgium. *Journal of Quaternary Science*, *17*(5–6), 547–559. doi:10.1002/jqs.707

Belgian Federal Government. (2011). *Kerncijfers landbouw 2011*. Brussels: De landbouw in België in cijfers.

Belgian Federal Government. (2013). *Kerncijfers landbouw 2013*. Brussels: De landbouw in België in cijfers.

Belgian Federal Government. (2014). *A Marine Spatial Plan for the Belgian part of the North Sea*. Brussels: De landbouw in België in cijfers.

Biswas, A. K. (2004). Response to Comments by Mitchell, Lamoree, and Dukhovny. *Water International*, *29*(4), 531–533. doi:10.1080/02508060408691818

Bourgeois, G. (2009). *Beleidsnota Toerisme 2009–2014*. Brussels: Vlaamse Regering.

Broothaers, L. (1995). *Geologie van Vlaanderen: een schets*. Brussels: Ministerie van de Vlaamse Gemeenschap, Afdeling Natuurlijke Rijkdommen en Energie.

Brouwers, J. et al. (2009). Chapter 11 'Climate Change and Water Systems'. In M. Van Steertegem (Ed.), *Flanders environmental outlook 2030*. Aalst: VMM.

Bursens, K., & Puype, S. (2011). *Stroomgebiedbeheerplannen voor Schelde en Maas: Naar een goede toestand van het watersysteem in Vlaanderen*: CIW.

Charlier, R. H., & De Meyer, C. P. (1992). Tourism and the coastal zone: The case of Belgium. *Ocean & Coastal Management, 18*(2–4), 231–240. doi:10.1016/0964-5691(92)90026-H

CIW. (2003). Decreet Integraal Waterbeleid

Colten, C. E., & Sumpter, A. R. (2009). Social memory and resilience in New Orleans. [Article]. *Natural Hazards, 48*(3), 355–364. doi:10.1007/s11069-008-9267-x

Cook, B. R., & Spray, C. J. (2012). Ecosystem services and integrated water resource management: Different paths to the same end? *Journal of Environmental Management, 109*, 93–100. doi:10.1016/j.jenvman.2012.05.016

Coudenys, H. et al. (2014). Sociale & economische omgeving. In A. K. Lescrauwaet, et al. (Eds.), *Compendium voor Kust en Zee 2013* (pp. 11). Belgium: Oostende.

Council Directive (EC) 76/160/EEC of 8 December 1975 concerning the quality of bathing water.

Council Directive (EC) 91/271/EEG of 21 May 1991 concerning urban waste-water treatment.

Council Directive (EC) 2000/60/EG of 23 October 2000 establishing a framework for Community action in the field of water policy.

Crabbé, A. (2008). Integraal waterbeleid in Vlaanderen: van fluïde naar solide. (Dissertation). Universiteit Antwerpen, Antwerp.

Crabtree, P. J. (2010). Agricultural innovation and socio-economic change in early medieval Europe: Evidence from Britain and France. *World Archaeology, 42*(1), 122–136. doi:10.1080/00438240903430373

De Moel, H. et al. (2012). Methods for future land-use projections for Flanders. Valorisation Report 2: CcASPAR.

De Sutter, R. (2011). Integrated Assessment of Spatial Climate Change Impacts in Flanders - Mirrored to the Dutch experiences. Valorisation Report 1: CcASPAR.

Declercq, G. (2000). De kustvlakte en de ontwikkeling van het graafschap Vlaanderen. *Vlaanderen: Tweemaandelijks Tijdschrift voor Kunst en Cultuur, 49*(3), 148–151.

EEA. (2006). *The changing faces of Europe's coastal areas*. Copenhagen: European Environment Agency.

Ervynck, A. et al. (1999). Human occupation because of a regression, or the cause of a transgression? *Probleme der Küstenforschung im südlichen Nordseegebiet, 26*, 97–121.

FAO. (2003). *Review of world water resources by country*. Rome: Food and Agriculture organization of the United Nations.

Flemish Government. (2012). *Vlaamse Baaien: Naar een geïntegreerde visie voor de kust*. Brussels: Fernand Desmyter.

Gerritsen, H. (2005). What happened in 1953? The big flood in the Netherlands in retrospect. *Philosophical Transactions of the Royal Society A: Mathematical, Physical and Engineering Sciences, 363*(1831), 1271–1291. doi:10.1098/rsta.2005.1568

Gleick, P. H. (2000). A look at twenty-first century water resources development. *Water International, 25*(1), 127–138. doi:10.1080/02508060008686804

Gobyn, R. (1987). Te kust en te kuur: badplaatsen en kuuroorden in België 16de-20ste eeuw.

GWP. (2000). *Integrated Water Resources Management Technical Advisory Committee Background Paper Number 4*. Stockholm: Global Water Partnership.

GWP. (2014). History of the Global Water Partnership. Retrieved from http://www.gwp.org/en/About-GWP/History/

Hartmann, T., & Tejo, S. (2014). Frontiers of land and water governance in urban regions. *Water International, 39*(6), 791–797. doi:10.1080/02508060.2014.962993

Heylen, J. (1997). De hydrologie van het IJzerbekken. *Water: Tijdschrift over Waterproblematiek, 16*(97), 239–244.

IMCORE. (2009). Case study: The Belgian coast. WP 2.5. In M. Willekens & F. Maes (Eds.): Innovative Management for Europe's Changing Coastal Resource.

IPCC. (2000). IPCC Special Report: Emissions scenarios.

IPCC. (2014). [Final draft] Chapter 23: Europe Climate Change 2014: Impacts, Adaptation, and Vulnerability. Working Group II IPCC.

Jonkman, S. N., & Vrijling, J. K. (2008). Loss of life due to floods. *Journal of Flood Risk Management, 1*(1), 43–56. doi:10.1111/j.1753-318X.2008.00006.x

Kellens, W., Zaalberg, R., Neutens, T., Vanneuville, W., De Maeyer, P. (2011). An analysis of the public perception of flood risk on the Belgian Coast. *Risk Analysis, 31*(7), 1055–1068. doi:10.1111/j.1539-6924.2010.01571.x

Koks, E. E., de Moel, H., & Bouwer, L. M. (2012). Effect of spatial adaptation measures on flood risk in the coastal area of Flanders. Valorisation Report 10b: CcASPAR.

Lebbe, L., Van Meir, N., & Viaene, P. (2010). Potential implications of sea-level rise for Belgium.

Loveluck, C., & Tys, D. (2006). Coastal societies, exchange and identity along the Channel and southern North Sea shores of Europe, AD 600–1000. *Journal of Maritime Archaeology, 1*(2), 140–169. doi:10.1007/s11457-006-9007-x

MDK. (2014, July 12). Onderlinge ligging van enkele vergelijkingsvlakken Retrieved from http://www.vlaamsehydrografie.be/Uplfiles/file/vergelijkingsvlakken.pdf

Mertens, T. et al. (2011). An integrated master plan for Flanders future coastal safety.

Mitchell, B. (2005). Integrated water resource management, institutional arrangements, and land-use planning. *Environment and Planning A, 37*(8), 1335–1352. doi:10.1068/a37224

Monballyu, M., Van den Driessche, E., & Pirlet, H. (2014). Toerisme & recreatie. In A. K. Lescrauwaet, et al. (Eds.), *Compendium voor Kust en Zee 2013* (pp. 10). Belgium: Oostende.

Nicholls, S., & Amelung, B. (2008). Climate change and tourism in Northwestern Europe: Impacts and adaptation. *Tourism Analysis, 13*(1), 21–31. doi:10.3727/108354208784548724

PBL. (2011). *Een delta in beweging. Bouwstenen voor een klimaatbestendige ontwikkeling van Nederland.* The Hague: Planbureau voor de Leefomgeving.

Rahaman, M. M., Varis, O., & Kajander, T. (2004). EU water framework directive vs. integrated water resources management: The seven mismatches. *International Journal of Water Resources Development, 20*(4), 565–575. doi:10.1080/07900620412331319199

Ruyck, A., Ampe, C., & Langohr, R. (2001). Management of the Belgian coast: Opinions and solutions. *Journal of Coastal Conservation, 7*(2), 129–144. doi:10.1007/BF02742475

Soens, T. (2006). Explaining deficiencies of water management in the late medieval Flemish coastal plain, 13th-16th centuries. *Jaarboek voor Ecologische Geschiedenis, 2005.*

Soens, T. (2009). *De spade in de dijk? Waterbeheer en rurale samenleving in de Vlaamse kustvlakte (1280–1580).* Gent: Academia Press.

Soens, T. (2011). Floods and money: Funding drainage and flood control in coastal Flanders from the thirteenth to the sixteenth centuries. *Continuity and Change, 26*(3), 333–365. doi:10.1017/S0268416011000221

SVW. (2008). *Strategisch plan drinkwatervoorziening Vlaanderen* (pp. 103). Antwerp: Samenwerking Vlaams Water.

Tempels, B., & Hartmann, T. (2014). A co-evolving frontier between land and water: Dilemmas of flexibility vs. robustness in flood risk management. *Water International, 39*(6), 872–883. doi:10.1080/02508060.2014.958797

Thoen, E., & Soens, T. (2003). Waterbeheer in de Vlaamse kustvlakte in de Late Middeleeuwen en het Ancien Régime: van landschapsgeschiedenis naar ecologische geschiedenis. *Jaarboek voor Ecologische Geschiedenis, 2001.*

Tys, D. (2005). De inrichting van een getijdenlandschap. De problematiek van de vroegmidde-leeuwse nederzettingsstructuur en de aanwezigheid van terpen in de kustvlakte: het voorbeeld van Leffinge. *Archeologie in Vlaanderen, 8,* 257–279.

Valcke, L. (1992). Van 'Woaterhuus' tot 'Woatertorre'. Deel 1: 1600–1855: waterputten en water-huis. *Gidsenblad Lange Nelle, 4,* 92–97.

Valcke, L. (1993). Van 'Woaterhuus' tot 'Woatertorre'. Deel 5: 1905–1925: Aanvoer en distributie van Bocq-water. *Gidsenblad Lange Nelle, 4,* 128–131.

Valcke, L. (1994). Van 'Woaterhuus' tot 'Woatertorre'. Deel 6: Na 1925: Verdere uitbreiding en automatisatie van het distributienet-4de watertoren. *Gidsenblad Lange Nelle, 1,* 3–8.

Van den Berghe, K., & De Sutter, R. (2014). De hydrologische cyclus in de Belgische kustzone en -polders. *Ruimte & Maatschappij, 5*(4), 32.

Van den Eynde, D., De Sutter, R., & Haerens, P. (2012). Evolution of marine storminess in the Belgian part of the North Sea. *Natural Hazards and Earth System Sciences, 12*(2), 305–312. doi:10.5194/nhess-12-305-2012

Van Passel, S., Nevens, F., Van Huylenbroeck, G., & Mathijs, E. (2005). Efficiëntie en Productiviteit van de Vlaamse landbouw: Een empirische analyse: Vlaamse Overheid: Steunpunt beleidsrele-vant onderzoek duurzame landbouw.

Verhulst, A. (2000). Historische ontwikkeling van het kustlandschap. *Tijdschrift voor Kunst en Cultuur, 49*(3), 7–10.

Vlaamse Baaien, T. H. V. (2014). Vlaamse Baaien: Veilig, natuurlijk, aantrekkelijk, duurzaam, ontwikkelend (pp. 98). Oostende.

VMM, & CIW. (2009). Strategische Visie Watervoorziening en Watergebruik.

VMM. (2006). Grondwaterbeheer in Vlaanderen: het onzichtbare water doorgrond (pp. 150).

VMM. (2008a). Het bekkenbeheerplan van het bekken van de Brugse Polders 2008-2013, Integraal waterbeleid in de praktijk: Integraal Waterbeleid, bekken van de Brugse polders.

VMM. (2008b). Het bekkenbeheerplan van het IJzerbekken 2008-2013, Integraal waterbeleid in de praktijk: Integraal Waterbeleid, bekken van de IJzer.

WES. (2009). Economische betekenis agrocomplex in West-Vlaanderen: eindrapport.

Westtoer. (2013). Trendrapport KiTS Kust 2007–2012.

A co-evolving frontier between land and water: dilemmas of flexibility versus robustness in flood risk management

Barbara Tempels[a] and Thomas Hartmann[b]

[a]Centre for Mobility and Spatial Planning, Ghent University, Ghent, Belgium; [b]Department of Human Geography & Regional Planning, Utrecht University, Utrecht, the Netherlands

Floods cause enormous damage on land and thus question the boundary between land and water in an extreme way. As floods increase in frequency and intensity, flood risk management must change from a resistance-based approach to a resilience approach. Whereas land uses require robust boundaries between land and water, the changing water system demands more flexible boundaries. This contribution discusses this tension from a theoretical perspective of resilience and co-evolution, using a socio-ecological systems approach. This offers a new perspective on the co-evolving frontier between land and water.

Introduction

River floods are among the most prominent, urgent and devastating consequences of climate change that one can experience in Europe. Climate change will increase both their intensity and likelihood in future (IPCC, 2014). This will affect urban areas in particular, because they are often located close to rivers or coastlines, thereby exposing valuable and vulnerable land uses to floods.

Traditionally, floods have been controlled with technical infrastructures (i.e. dikes and dams) (Patt & Juepner, 2013). Despite major investments in such flood protection measures (Loucks et al., 2008), the annual damage increased over the past decades (Munich Re, 2010), suggesting that this approach might no longer effectively reduce flood risks. Urban developments in floodplains contribute to the problem in two ways: first, space for the rivers shrinks and water levels increase downstream; and second, most settlements are not adapted to inundations, exposing people and assets to floods (Hartmann, 2011b; Patt & Juepner, 2013; Petrow et al., 2006). If no other approach to flood risk management is chosen, this entrenches a lock-in situation in technical flood protection approaches because existing settlements can hardly be removed (Hartmann, 2011a).

In fact, in recent decades, new approaches in dealing with floods have been discussed in the literature and in practice. Flood policy is shifting from the rather robust defence against floods towards a more flexible and adaptive flood risk management (Hartmann & Juepner, 2014; Patt & Juepner, 2013). This shift questions established physical and governance boundaries between land and water. Whereas traditional approaches promote robust boundaries between dry and wet land, adaptive approaches introduce a fluid frontier between the two.

The more flexible flood risk management approach conflicts with the robustness of existing spatial structures and land-use rights. This tension is an important reason why more flexible and adaptive approaches are not yet standard practice. However, insights are lacking on how to balance the simultaneous but conflicting needs for flexibility and robustness of this boundary.

This paper identifies key questions of the dilemma of flexibility versus robustness in flood risk management. It therefore sets an agenda for future research. The tension between flexibility and robustness is discussed from a theoretical perspective. This discussion is supported by general observations and examples from flood risk management practice, derived from previous work by the authors and the literature (Hartmann, 2011a; Tempels, 2013). Using a socio-ecological systems approach, a new perspective on the frontier between land and water is developed, based on resilience and co-evolutions between land and water. By reframing the issue in a complex adaptive systems approach of co-evolution between water and land, the governance of flood risks can become more effective.

From flood protection to flood risk management

In principle, floods can be approached with two different concepts: increasing the robustness, or accepting the risk and adapting to it (flexibility). The first usually requires modelling and prediction, technical flood protection measures such as dikes, and strong water management institutions with technical skills. The latter depends on comprehensive and integrative concepts, encompassing many stakeholders and asking for collaboration at various levels. Adaptability does not mean just amending the city, thus enabling the existing urban structure to remain the same. Rather, adaptive cities will become trans-formed by (the threat of) flood events.

Flood protection

Since the beginning of industrialization, flood protection has been the dominant approach in most European countries. It is based on the assumption that floods are predictable, with a more or less constant trend in the flooding frequency. Subsequently they may be constrained through engineered solutions (Fleming, 2002; Johnson & Priest, 2008; Patt & Juepner, 2013) and floodplains can be made available for all kinds of land uses (Hartmann, 2011b; Loucks, 2000). In this approach, emphasis is on absorbing shocks, limiting short-time damage, performing a speedy recovery back to the same functions. The goal is to preserve existing developments by defending against the water and enforcing a strong boundary between land and water (Hartmann, 2009).

The advantage of flood protection is that it enables constant conditions for settlements behind the dikes, and therefore facilitates using (protected) land efficiently without making compromises because of a flood risk. Resistance is easier to live with in everyday life. It enables easier decision-making for land-use planners and clear division of respon-sibilities between water management and spatial planning (Hartmann & Driessen, 2014).

Flood risk management

In contrast to flood protection, flood risk management does not mean the quest for fail-safe options to prevent flooding. It rather assumes that flood risks vary and calamities will happen. Flood risk management asks for adaptations of vulnerable objects in order to minimize the consequences of floods, but at the same time it allows some flooding (Vis et al., 2003). This vulnerability encompasses not only (infra)structural aspects, but also social aspects such as adaptive capacities, which determine communities' ability to cope with flooding. Examples for physically resilient structures include floating homes (Pierdolla, 2008) and adapted interiors for houses (e.g. not putting electrical installations in the basement), but also escape routes for evacuations or calamity polders (Internationale Kommission zum Schutz des Rheins (IKSR), 2002), and even, in some cases, abandoning certain areas (McLeman & Smit, 2006).

In addition to adjustments and restructuring of physical structures, also the socio-economic and political setting of flooding needs to be examined. Adaptive capacities are a result of several social, economic, technological, knowledge-related, institutional and cultural mechanisms (Brouwer et al., 2007). However, these mechanisms and their inter-actions are very complex, making increasing adaptive capacities less straightforward. It involves financial recovery capacity, insurance schemes (Berke & Campanella, 2006; Clark, 1998), liability issues, availability of information, etc.

These examples show that resilience comes with costs for adaptation and compromises for land uses. In addition, it challenges existing institutions and well-entrenched modes of governance (van den Brink, 2009).

The list of examples also reveals that centralized governmental institutions such as water management agencies can hardly implement flood risk management on their own. Flood risk management asks for the compliance and cooperation of not only many different institutions, but also of public and private stakeholders (Loucks et al., 2008). So, not only does flood risk management require a fundamental rethinking of existing working paradigms within water management agencies, but also this shift of paradigms needs to be supported and sustained by various stakeholders with sometimes competing interests: public and private actors, comprehensive and sectoral planning, central and decentralized structures. A new mode of governance that balances these issues of flex-ibility and robustness is needed. Before discussing the relevant questions, we need to elaborate where the needs for flexibility and robustness are coming from.

Flexibility versus robustness

The turn from flood protection to flood risk management is triggered by a need for flexibility due to changing conditions. But where does this need for flexibility and robustness come from? In what follows, the context that shapes these simultaneous but contrasting needs is drawn.

The need for flexibility

The water system is influenced by complex natural–physical components (Patt & Juepner, 2013). For example, the exact occurrence and intensity of climate extremes is unpredict-able in the long-term, as the climate is inherently variable. Moreover, the climate seems to be changing towards an increasing intensity and frequency of flooding (IPCC, 2014).

Additionally, human interventions induce (intentional or unintentional) alterations to the water system. Technical infrastructures such as dikes and dams, upstream activities, and also land uses in the catchment have considerable impacts on the water system. Particularly in urban areas of developed countries, the multiple and intense land-use activities in catchments make the prediction and management of the water system more challenging and complex. For example, the urbanization of floodplains takes up space for the rivers and also increases discharge of rainwater due to sealed surfaces. Urbanization also creates local heat islands with their own microclimate, making the flood forecast more difficult.

But also social aspects of flood management are subject to long-term change. Considering multiple actors (water managers, politicians, residents, etc.) leads to relational uncertainty (Brugnach et al., 2008). This type of uncertainty emerges from the parallel and equivalent existence of multiple knowledge frames. Different actors understand the issue differently, hold different values and beliefs and therefore have different judgments about the potential actions or interventions. Therefore, decision-making is characterized by uncertainty regarding the outcome of the decision (Tompkins & Adger, 2004).

All these elements are associated with a range of uncertainties (Dessai & van der Sluijs, 2007) and complexities that cannot be mitigated through modelling or further research, as they are inherently unpredictable. Therefore, flood management strategies can no longer be based on the conventional linear methods of risk assessment, which evaluate alternative measures to implement the optimal solution. The inherent uncertainty and associated complexity with respect to changes in the physical and social components of flood risk require more flexible schemes to be incorporated into decision processes and management choices.

However, there are some clear disadvantages and discomforts to more adaptable approaches, such as physical constraints to remove built structures or high costs (monetary compensations) and social difficulties (issues of justice, legal certainty and liability) when changing land-use allocations. Nevertheless, there is no adequate alternative: facing increasing floods and continuing urban developments in flood-prone areas, traditional approaches to floods fail, and flexibility is becoming an essential component of future flood risk management.

The demand for robustness

An important argument for traditional flood protection is that it provides a robust setting for all kinds of activities behind dikes. This goes back to the pioneers of water engineering (Nisipeanu, 2008). Building a dike along a river essentially increases the value of property rights behind the dike, because the land becomes attractive for building activities. Spatial planning decisions in those areas are based on the assumption that a certain piece of land remains physically consistent over a long period of time. Changing such a designation is rather difficult (as discussed above). Needham and Hartmann (2012) conclude that property rights are inevitable but also desirable: they are inevitable because whenever a spatial plan or a planning measure specifies how a particular plot of land may or may not be used, it is socially constructing and assigning property rights through the law; they are desirable because property rights make planning decisions robust. So, robust planning decisions are essential for the functioning of society – the whole system of property rights, and thus economic investment, builds on reliable and robust spatial planning decisions.

However, contemporary planning theory often criticizes such property-oriented spatial planning as being too inflexible to cope with uncertainties and wicked situations (Bertolini, 2010; Hartmann et al., 2012a). Planning thus creates lock-in situations. A spatial allocation and distribution of goods that might have been desirable at one time can become inconvenient or even dangerous, as seen in the case of riparian urban development and increasing floods. Moreover, planning theory asserts that 'in the everyday world of spatial planning practice, planners are more likely to rely on intuition or practical wisdom' (Hillier, 2010, p. 11). To some extent, planners guess (Paterson, 2007) and experiment (Bertolini, 2010; Hillier, 2010) with space. However, abandoning robust spatial planning decisions and changing towards a system based entirely on flexibility is also not an option. The robustness of spatial planning decisions is and remains an essential element for the functioning of our society.

The concept of resilience: balancing flexibility and robustness

So on the one hand there is a need for robustness within planning, while on the other hand there is also a need for flexibility emanating from changing flood risks. Both claims are legitimate (Needham & Hartmann, 2012), and both the approaches have advantages and disadvantages (Table 1). Therefore, it is not a question of choosing one above the other; rather, the question is how to accommodate both needs. Therefore, a new balance between flexibility and robustness needs to be found to govern land and water effectively in urban areas. The rest of this paper addresses this balance by discussing the resilience approach and the co-evolutionary interactions between flood risk management and society.

Resilience is often discussed as a new flood management approach (Begum et al., 2007; Bruijn, 2005; Petrow et al., 2006; Roth & Warner, 2007). Although resilience is best known as an advocate for more flexible approaches, the concept also stresses the

Table 1. Flood protection and flood risk management.

	Flood protection (robustness)	Flood risk management (flexibility)
Perception of flooding	Floods are predictable, with a more or less constant trend in flooding frequency	Flood risks vary and are unpredictable
Perception of damage	Quest for fail-safe options	Calamities will happen
Goal	Preservation oriented	Allow for reorganization and development, enable the system to adapt to changing conditions
Means	Defending against the water and enforcing a strong boundary between land and water	Adaptation of vulnerable objects to minimize the consequences of floods, but also allows some flooding
Advantages	Constant conditions: • Facilitates using (protected) land efficiently without compromises • Easier decision-making for land-use planners • Clear division of responsibilities between water management and spatial planning	Deals better with uncertainty and associated complexity with respect to changes in the physical and social components of flood risk
Disadvantages	Too inflexible to cope with uncertainties and change May create lock-in	Costs for adaptation and compensation Compromises for land uses Issues of justice, legal certainty and liability Challenges existing institutions and well-entrenched modes of governance Compliance and cooperation of not only many different institutions, but also of public and private stakeholders

need to balance robustness and flexibility. On a very basic level, resilience describes the ability of a system to absorb disturbances (shocks); so it means that cities are, one way or another, able to absorb the negative consequences of flooding. In this view, it advocates a more flexible approach as a response to the changing conditions in flood risks (e.g. climate change and socio-economic developments), while retaining some robustness. In fact, earlier conceptualizations of resilience (i.e. engineering and ecological resilience) mainly focused on maintaining stability and being persistent or robust within certain boundaries against disturbances. However, more recent interpretations challenge this equilibristic view (Davoudi et al., 2012). Based on coupled socio-ecological systems, the importance of renewing, regenerating, and reorganizing following a disturbance is emphasised. In other words, the resilience concept encompasses both being persistent or robust (robustness), and at the same time being able to renew, regenerate and reorganize (flexibility).

Nevertheless, focus in practice has been more on bouncing back and short-term damage reduction (robustness), and less on the capacity for reorganization and development (flexibility) (Folke, 2006). To overcome this, Davoudi et al. (2012) propose an evolutionary approach where long-term change is necessary in the face of changing conditions. Resilience is then 'not conceived of as a return to normality, but rather as the ability of complex socio-ecological systems to change, adapt, and, crucially, transform in response to stresses and strains' (p. 302).

But how does the concept of resilience add understanding to how to deal with the tension between flexibility and robustness in flood risk management? The theory of resilience is based on socio-ecological systems. The idea that social and ecological systems develop in co-evolution with each other can add some perspective.

Co-evolution between social and natural systems

Traditionally, floods are framed as purely natural–physical disturbances in the water system. As such, they are external threats to human systems. By framing floods like this, solutions are usually confined within the boundaries of the water system and water management, and intended to minimize and, if possible, even eliminate floods. However, as indicted above, socio-spatial aspects (e.g. vulnerable urban developments in flood-prone areas or settlements in potential retention areas upstream) also substantially contribute to flood risks, i.e. both the probability of flooding and potential flood losses. Taking this into account, on the one hand, charges flood risks with additional complexity, but also implies that potential solutions can also be found in socio-spatial interventions, e.g. by lowering vulnerabilities. So the issue of flooding rests at the intersection of the water system (water flows, engineering infrastructures etc.) and the socio-spatial system (settlements and spatial development). Consequently, integrating socio-spatial systems in flood risk management can lead to more comprehensive view on the issue.

Considering floods as a result of the interaction of social and physical systems sheds a new light on flood management (Gerrits, 2008). This perspective is called 'co-evolution'. Kallis (2007, p. 4) states that 'a co-evolutionary explanation [...] entails two or more evolving systems whose interaction affects their evolution'. Floods are inextricably results of co-evolving land (socio-spatial) and water (natural–physical) systems (Folke et al., 2002; Tompkins & Adger, 2004). This means that flood risks influence land-use options, and socio-spatial developments on land in turn have an impact on flood risks (e.g. increased run-off) (Gerrits, 2011; Hartmann, 2010; Mitleton-

Kelly, 2003). The mechanisms behind spatial developments respond to (changes in) flood risks (Hartmann, 2011a; Pahl-Wostl, 2006). These include spatial demands, real estate markets, insurance systems (Botzen et al., 2009), knowledge of flood risks (Bubeck et al., 2012), perceptions and attitudes towards floods, and the behaviour and practices of the broader society. The presence of valuable spatial developments in flood-prone areas, on the other hand, causes a need for protection through technical infra-structure, governmental rules, engineering rules and technology. Finally, co-evolution provides an analytical framework to understand the interdependent evolution of social and environmental subsystems.

An example for such co-evolution of boundaries between land and water can be found in Nijmegen in the Netherlands. The 'Waalsprong' is a huge urban expansion project north of the centre of Nijmegen, across the River Waal. The project is part of the 'Room for the River' programme initiated by the Dutch government, which combines water safety targets with spatial planning goals (Coninx & Cuppen, 2010). At the point where the development is occurring, the Waal bends sharply and also becomes narrower. In 1993 and 1995 this location was subject to flooding. The extension plans already existed before the Room for the River programme. However, the programme added that the urban development and flood protection support each other. The chosen solution is to move the existing Waal dike in Lent a few hundred metres inland to restore the river's floodplain and to construct an ancillary channel there. This enables the hinterland to develop while at the same time preparing a sufficient buffer for flood risks. This provides both robustness (the dike) and flexibility (the creation of a floodplain and an ancillary channel), enabling the frontier between land and water to co-evolve.

Also in the urban regeneration project HafenCity in the centre of Hamburg, Germany, similar considerations were taken into account. This site is located outside of Hamburg's main dike line, and is hence prone to flooding. All roads and bridges were elevated to the minimum height corresponding to the flood walls protecting the inner city, while the bases of the buildings were constructed so that they are flood secure. Instead of altering the water system, adjustments in the spatial system were implemented to allow flooding. Although this approach is still quite technical and engineered, it reflects a shift towards accommodating more flexibility.

The lack of co-evolution in flood risk management

In the traditional robustness-based approach to floods, this co-evolution is not acknowl-edged. When framing the flooding issue as a purely physical problem (as discussed above), the societal context (including spatial developments) is seen as being external and unalterable, enabling and restricting flood management options (Hutter, 2006). The interaction between land and water is then one-directional: what happens on the land has consequences for the management of the water system, but land uses rarely respond to changes in the water system (Figure 1). This traditional static conceptualization of the societal context does not reflect the dynamic and reciprocal co-evolution of both systems (Boisot & Child, 1999).

Co-evolution and flexibility versus robustness

The examples illustrate how a co-evolutionary perspective to the two systems of land and water can help in finding a new balance between robustness and flexibility in flood risk management. Co-evolution is more than the mutual influence between both systems

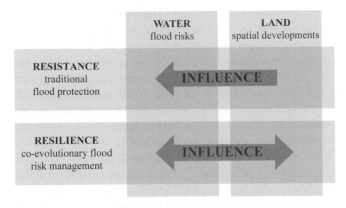

Figure 1.   Interactions between land and water in flood protection and resilience approaches.

(Gerrits, 2011). In order for co-evolution to be fruitful, it is important that both systems are dynamic.

A co-evolving system tries to adapt to the environment when necessary, and it tries to influence its environment when possible (Edelenbos & van Buuren, 2006). The discussion of flexibility versus robustness thus comes down to accommodating both changing flood risks (when necessary) and stable social development by influencing the water system (when possible) in a co-evolutionary process. In the case of flood management, this means that spatial planning and water management need to be adapted to each other. Currently, in practice, there is a tendency towards this approach due to the increasing importance of spatial planning within the flooding issue (e.g. Coninx & Cuppen, 2010) and the growing interest in co-evolutionary planning (Boelens & De Roo, forthcoming).

Discussion and conclusions: co-evolution as a perspective that bridges the socio-spatial and the natural–physical system

So, on the one hand, we have a social system that is asking for robustness, while, on the other hand, changes in the water system demand flexibility. To understand this tension between flexibility and robustness, the concept of co-evolution between land (socio-spatial systems) and water (as a natural–physical system) is proposed. Actions on land affect the water system, while flood risks emanating from the water system affect spatial development options. This can help in framing the flooding issue more realistically, as far as interactions between flood management choices and society are considered. For example, flood risks are generally defined as the probability of flooding weighed against the potential damage. These two variables are often treated as independent variables; flood management strategies lower either the probability of flooding or the damage in case of flooding. However these two variables are dependent, arising from the mutual influences between the water and land system. Measures to lower the probability of flooding (e.g. building dikes) influence the development of potential damage (e.g. construction of new buildings). Vice versa, the presence of vulnerable groups or structures heightens the need for protection from flooding.

Currently, there are few insights into these interactions. On the one hand, social research describes issues of vulnerability (Grothmann & Reusswig, 2006; Siegrist & Gutscher, 2008), while hydrological models estimate the effects of infrastructural interventions on the probability of flooding. Therefore, the interactions between flood management choices and society need to be analysed and monitored.

The concept of co-evolution does not per se provide a solution to the dilemma of flexibility versus robustness, but it offers another perspective that that bridges the socio-spatial demand for robustness and the natural–physical constraint and need for flexibility in the interplay of land and water at its fluid frontier. By understanding the mutual influence (the co-evolutionary character) of the two systems, the perspective on flood risk management measures changes. When drafting measures in one system, the effects on the other system should be considered so as to obtain a more realistic estimate of the resulting flood risks. When areas are protected from flood risks, what does this imply in terms of spatial development perspectives? What are the effects on flood risks of urban development, locally and also downstream? And what does this mean in the long term? By considering the interactions and co-evolutionary nature of 'water' and 'land' systems, more comprehensive and effective results can be expected.

Although this seems to be obvious, this is not yet standard practice. Often, the focus in flood risk management is more on the water system and less on the effects of the land system and how they influence water issues. Flood risk management measures are generally restricted within the boundaries of the water system, while within spatial planning, a remedial approach to managing flood risks is used.

Finally, this contribution offers not more and not less than a perspective on the dilemma of balancing robustness and flexibility in flood risk management. The co-evolutionary perspective, discussed above, raises a couple of essential and important research questions. One of the issues has to do with the costs of flexibility (adaptation measures, but also compensation claims for disturbing the robust system 'land'). Also, questions of justice and equity need to be dealt with: if flood risk management requires a more flexible approach to floods, who will get what kind of protection? This requires new discourses on the risk absorption capacities of land uses; but it also raises a couple of legal issues related to liabilities or responsibilities. Further attention needs to be paid to these questions. The perspective of co-evolution helps understanding the interdependencies of the social and environmental subsystems land and water – thus helping understanding the fluid frontier between the two.

Funding

This paper draws from research conducted within the Policy Research Centre on Spatial Development, funded by the Flemish Government (Belgium).

References

Begum, S.Stive, M. J. F. & Hall, J. W. (Eds). (2007). *Flood risk management in Europe: Innovation in policy and practice*. Dordrecht: Springer.

Berke, P. R., & Campanella, T. J. (2006). Planning for postdisaster resiliency. *The ANNALS of the American Academy of Political and Social Science, 604*(1), 192–207. doi:10.1177/0002716205285533

Bertolini, L. (2010). Complex systems, evolutionary planning? In G. de Roo & E. A. Silva (Eds.), *A planner's encounter with complexity* (pp. 81–98). Farnham, Surrey: Ashgate.

Boelens, L., & De Roo, G. (forthcoming). Planning of undefined becoming; From post-structuralism towards actor-relational opportunities. *Planning Theory*. doi:10.1177/1473095214542631

Boisot, M., & Child, J. (1999). Organizations as adaptive systems in complex environments: The case of China. *Organization Science, 10*(3), 237–252. doi:10.1287/orsc.10.3.237

Botzen, W., Aerts, J., & van den Bergh, J. (2009). Willingness of homeowners to mitigate climate risk through insurance. *Ecological Economics, 68*(8–9), 2265–2277. doi:10.1016/j. ecolecon.2009.02.019

Brouwer, R., Akter, S., Brander, L., & Haque, E. (2007). Socioeconomic vulnerability and adaptation to environmental risk: A case study of climate change and flooding in Bangladesh. *Risk Analysis, 27*(2), 313–326. doi:10.1111/j.1539-6924.2007.00884.x

Brugnach, M., Dewulf, A., Pahl-Wostl, C., & Taillieu, T. (2008). Toward a relational concept of uncertainty: About knowing too little, knowing too differently, and accepting not to know. *Ecology and Society, 13*(2), 30. Retrieved from http://www.ecologyandsociety.org/vol13/iss2/art30/

Bruijn, K. M. de. (2005). *Resilience and flood risk management: A systems approach applied to lowland rivers.* Delft: Delft University Press.

Bubeck, P., Botzen, W. J. W., & Aerts, J. C. J. H. (2012). A review of risk perceptions and other factors that influence flood mitigation behavior. *Risk Analysis, 32*(9), 1481–1495. doi:10.1111/j.1539-6924.2011.01783.x

Carpenter, S., Walker, B., Anderies, J. M., & Abel, N. (2001). From metaphor to measurement: Resilience of what to what? *Ecosystems, 4*(8), 765–781. doi:10.1007/s10021-001-0045-9

Clark, M. J. (1998). Flood insurance as a management strategy for UK coastal resilience. *The Geographical Journal, 164*(3), 333. doi:10.2307/3060621

Coninx, I., & Cuppen, M. (2010). *Institutionele antwoorden op complexiteit: een vergelijkende analyse van hoogwaterbeleid/overstromingsbeheer in de 'lage landen'.* Paper presented at the Politicologenetmaal 2010, Leuven, Belgium.

Davoudi, S., Shaw, K., Haider, L. J., Quinlan, A. E., Peterson, G. D., Wilkinson, C., . . . Davoudi, S. (2012). Resilience: A bridging concept or a dead end? 'Reframing' resilience: Challenges for planning theory and practice interacting traps: Resilience assessment of a pasture management system in Northern Afghanistan Urban Resilience: What does it mean in planning practice? Resilience as a useful concept for climate change adaptation? The politics of resilience for planning: A cautionary note *Planning Theory & Practice, 13*(2), 299–333. doi:10.1080/14649357.2012.677124

Dessai, S., & van der Sluijs, J. (2007). *Uncertainty and climate change adaptation: A scoping study.* Utrecht: Copernicus Institute for Sustainable Development and Innovation, Department of Science Technology and Society.

Edelenbos, J., & van Buuren, M. W. (2006). Innovations in the Dutch Polder: Communities of practice and the challenge of coevolution. *Emergence: Complexity and Organization, 8*(1), 42–49. Retrieved from http://hdl.handle.net/1765/10654

Fleming, G. (2002). *Flood risk management: Learning to live with rivers.* London: Thomas Telford.

Folke, C. (2006). Resilience: The emergence of a perspective for social–ecological systems analyses. *Global Environmental Change, 16*(3), 253–267. doi:10.1016/j.gloenvcha.2006.04.002

Folke, C., Carpenter, S., Elmqvist, T., Gunderson, L., Holling, C. S., & Walker, B. (2002). Resilience and sustainable development: Building adaptive capacity in a world of transformations. *AMBIO: A Journal of the Human Environment, 31*(5), 437–440.

Gerrits, L. (2008). *The Gentle Art of Coevolution: A complexity theory perspective on decision making over estuaries in Germany, Belgium and the Netherlands.* Rotterdam: Erasmus University Rotterdam.

Gerrits, L. (2011). A coevolutionary revision of decision making processes: An analysis of port extensions in Germany, Belgium and the Netherlands. *Public Administration Quarterly, 35*(3), 309–341. Retrieved from http://www.jstor.org/stable/41506760

Grothmann, T., & Reusswig, F. (2006). People at risk of flooding: Why some residents take precautionary action while others do not. *Natural Hazards, 38*(1–2), 101–120. doi:10.1007/s11069-005-8604-6

Hartmann, T. (2009). Clumsy floodplains and the law: Towards a responsive land policy for extreme floods. *Built Environment, 35*(4), 531–544. doi:10.2148/benv.35.4.531

Hartmann, T. (2010). Reframing polyrational floodplains: Land policy for large areas for temporary emergency retention. *Nature & Culture, 5*(1), 15–30. doi:10.3167/nc.2010.050102

Hartmann, T. (2011a). *Clumsy floodplains: Responsive land policy for extreme floods.* Farnham, Surrey: Ashgate.

Hartmann, T. (2011b). Contesting land policies for space for rivers - rational, viable, and clumsy floodplain management. *Journal of Flood Risk Management, 4*(3), 165–175. doi:10.1111/j.1753-318X.2011.01101.x

Hartmann, T. (2012). Land policy for German Rivers: Making space for the rivers. In J. F. Warner, A. van Buuren, & J. Edelenbos (Eds.), *Making space for the river. Governance experiences with multifunctional river planning in the US and Europe* (pp. 121–133). London: IWA Publishing.

Hartmann, T., & Driessen, P. P. (2014). The flood risk management plan: Towards spatial water governance. *Journal of Flood Risk Management,* n/a. doi:10.1111/jfr3.12077

Hartmann, T., & Juepner, R. (2014). The flood risk management plan: An essential step towards the institutionalization of a paradigm shift. *International Journal of Water Governance, 2*, 107–118. doi:10.7564/13-IJWG5

Hartmann, T., & Jüpner, R. (2013). Der Hochwasserrisikomanagementplan – Herausforderung für Wasserwirtschaft und Raumplanung. In J. Stamm & K. -U. Graw (Eds.), *Wasserbaukolloquium 2013. Technischer und organisatorischer Hochwasserschutz - Bauwerke, Anforderungen, Modelle* (pp. 183–192). Dresden: Wasserbauliche Mitteilungen.

Hartmann, T., & Needham, B. (2012a). Introduction: Why reconsider planning by law and property rights? In T. Hartmann & B. Needham (Eds.), *Planning by law and property rights reconsidered* (pp. 1–23). Farnham, Surrey: Ashgate.

Hartmann, T. & Needham, B. (Eds.) (2012b). *Planning by law and property rights reconsidered.* Farnham, Surrey: Ashgate.

Hillier, J. (2010). Introduction: Planning at yet another crossroads? In J. Hillier & P. Healey (Eds.), *The Ashgate research companion to planning theory. Conceptual challenges for spatial planning* (pp. 1–34). Farnham, Surrey: Ashgate.

Holling, C. S. (1973). Resilience and stability of ecological systems. *Annual Review of Ecology and Systematics, 4*(1), 1–23. doi:10.1146/annurev.es.04.110173.000245

Holling, C. S. (1996). Engineering resilience versus ecological resilience. In P. C. Schulze (Ed.), *Engineering within ecological constraints* (pp. 31–43). Washington, D.C: National Academy Press.

Hutter, G. (2006). Strategies for flood risk management - A process perspective. *Flood Risk Management: Hazards, Vulnerability and Mitigation Measures, 67*, 229–246.

Internationale Kommission zum Schutz des Rheins. (2002). *Hochwasservorsorge Maßnahmen und ihre Wirksamkeit.* Retrieved from http://www.iksr.org/fileadmin/user_upload/Dokumente_de/RZ_iksr_dt.pdf

IPCC. 2014. Climate Change 2014: Impacts, Adaptation, and Vulnerability: IPCC WGII AR5 Chapter 23. IPCC. Retrieved from http://www.ipcc.ch/report/ar5/wg2/

Janssen, M. A., Schoon, M. L., Ke, W., & Börner, K. (2006). Scholarly networks on resilience, vulnerability and adaptation within the human dimensions of global environmental change. *Global Environmental Change, 16*(3), 240–252. doi:10.1016/j.gloenvcha.2006.04.001

Johnson, C. L., & Priest, S. J. (2008). Flood risk management in England: A changing landscape of risk responsibility. *International Journal of Water Resources Development, 24*(4), 513–525. doi:10.1080/07900620801923146

Kallis, G. (2007). Socio-environmental co-evolution: Some ideas for an analytical approach. *International Journal of Sustainable Development and World Ecology, 14*(1), 4–13. doi:10.1080/13504500709469703

Klein, R. J. T., Nicholls, R. J., & Thomalla, F. (2003). Resilience to natural hazards: How useful is this concept? *Environmental Hazards, 5*(1), 35–45. doi:10.1016/j.hazards.2004.02.001

Loucks, D. P. (2000). Sustainable water resources management. *Water International, 25*(1), 3–10. doi:10.1080/02508060008686793

Loucks, D. P., Stedinger, J. R., Davis, D. W., & Stakhiv, E. Z. (2008). Private and public responses to flood risks. *International Journal of Water Resources Development, 24*(4), 541–553. doi:10.1080/07900620801923286

McLeman, R., & Smit, B. (2006). Migration as an adaptation to climate change. *Climatic Change, 76*(1–2), 31–53. doi:10.1007/s10584-005-9000-7

Mitleton-Kelly, E. (2003). Complex systems and evolutionary perspectives of organisations: The application of complexity theory to organisations. In E. Mitleton-Kelly (Ed.), *Advanced series in management. Complex systems and evolutionary perspectives on organisations. The application of complexity theory to organisations* (pp. 23–50). Oxford: Emerald.

Munich Re (Ed). (2010). *Natural catastrophes 2010: Analyses, assessments, positions. Topics Geo, 2010.* Munich: Munich Reinsurance Company.

Needham, B., & Hartmann, T. (2012). Conclusion: Planning by law and property rights reconsidered. In T. Hartmann & B. Needham (Eds.), *Planning by law and property rights reconsidered* (pp. 219–227). Farnham, Surrey: Ashgate.

Nisipeanu, P. (2008). Tradition oder Fortentwicklung? Wasserrecht im UGB. *Natur und Recht, 30,* 87–97.

Pahl-Wostl, C. (2006). Transitions towards adaptive management of water facing climate and global change. *Water Resources Management, 21*(1), 49–62.

Paterson, J. (2007). Sustainable development, sustainable decisions and the precautionary principle. *Natural Hazards, 42*(3), 515–528.

Patt, H. & Juepner, R. (Eds). (2013). *Hochwasser Handbuch: Auswirkungen und Schutz.* Heidelberg: Springer.

Petrow, T., Thieken, A. H., Kreibich, H., Merz, B., & Bahlburg, C. H. (2006). Improvements on flood alleviation in Germany: Lessons learned from the Elbe flood in August 2002. *Environmental Management, 38*(5), 717–732.

Pierdolla, M. (2008). *Floating Homes: Eine schwimmende Wohnform als neues städtebauliches Element in Deutschland.* Saarbrücken: VDM Verlag Dr. Müller.

Roth, D., & Warner, J. (2007). Flood risk, uncertainty and changing river protection policy in the Netherlands: The case of calamity polders? *Tijdschrift voor Economische en Sociale Geografie, 98*(4), 519–525.

Siegrist, M., & Gutscher, H. (2008). Natural hazards and motivation for mitigation behavior: People cannot predict the affect evoked by a severe flood. *Risk Analysis, 28*(3), 771–778.

Tempels, B. (2013). Veerkracht En Ruimtelijke Planning. *Agora, 29*(4), 40–44.

Tompkins, E. L., & Adger, W. N. (2004). Does adaptive management of natural resources enhance resilience to climate change? *Ecology and Society, 9*(2), 1–14. Retrieved from http://www.ecologyandsociety.org/vol9/iss2/art10/

Turner, B. (2010). Vulnerability and resilience: Coalescing or paralleling approaches for sustainability science? *Global Environmental Change, 20*(4), 570–576.

van den Brink, M. (2009). *Rijkswaterstaat on the horns of a dilemma.* Delft: Eburon.

Vis, M., Klijn, F., Bruijn, K. de, & van Buuren, M. (2003). Resilience strategies for flood risk management in the Netherlands. *International Journal of River Basin Management, 1*(1), 33–40.

Urban planning lock-in: implications for the realization of adaptive options towards climate change risks

Karen Hetz[a,b] and Antje Bruns[a]

[a]Department of Geography, Humboldt Universität zu Berlin, Berlin, Germany; [b]School of Architecture and Planning, University of the Witwatersrand, Johannesburg, South Africa

Urban planning can play a potentially meaningful role in managing the risks of climate change. It is, however, unclear to what extent planning practice can be transformed in order to address these risks effectively in the global south. Using Johannesburg in South Africa as an illustrative case and the interrelated challenges of flood risks and informal growth as an example, it is demonstrated how the identification of a particular planning practice with historically informed values of justice substantially constrains the realization of adaptive planning options. Correspondingly, its implications for managing flood risks of climate change through planning under conditions of urban divide are outlined.

Introduction

Urban flooding poses various risks to many metropolises in the global south. These risks are projected to worsen due to increases in the magnitude and frequency of extreme precipitation events, a consequence of climate change (IPCC, 2014; Milly et al., 2002). This has negative implications in particular for already vulnerable urban populations, such as informal dwellers, and with respect to urban liveability: damage to property, health risks and fatalities as the direct result of flooding in informal settlements are well documented (Douglas, Alam, & Maghenda, 2009; White, 2012; Wisner, 2009). Informal settlements lack storm water infrastructure, have insufficient sanitary infrastructure and inappropriate solid waste management; hence, flooding events in these areas also result in further degradation of urban water quality as contaminated particles are washed out and inserted untreated into the cities' streams (Satterthwaite, Huq, & Reid, 2009).

Although many environmental risks, such as the risks of urban flooding, are exacerbated by climate change, they also relate back to products of informal or environmentally unconscious developments and unintended consequences of formal planning: 'The effects of urbanisation and climate change are converging in dangerous ways' (UN-Habitat, 2011a, p. 1). In the global south informal settlements are particularly prone to flooding since they are often located in risk areas such as flood plains (Figure 1) that are unattractive for formal developments, thus providing dwellers a certain degree of safety from evictions (Davis, 2007). Due to climate change, climate refugees from rural areas are

Parts of the research were conducted during membership of the corresponding author to the Dresden Leibniz Graduate School (DLGS).

Figure 1. Informal settlement in floodplains of the Jukskei River, Johannesburg. Source: Hetz (2011).

likely to contribute to further in-migration to metropolitan cities of the global south (UN-Habitat, 2011), with informal settlement serving as the common entry point for many rural–urban migrants.

Given pre-existing urban environmental risks that are projected to increase further due to climate change, effectively addressing climate change issues through planning practices consequently implies the realization of adaptive planning options, which better address current flood risks in informal urban development settings and proactively respond to anticipated informal growth and related vulnerabilities as well. The paper builds on recent publications that suggest that due to the interrelation of risks and urban development patterns, urban planning plays a potentially meaningful role in adapting cities' built environments to the effects of climate change (Bulkeley, 2013; UN-Habitat, 2011; Wilson & Piper, 2011), and in improved flood risk management (Davar et al., 2001). Urban planning thereby offers a long-term perspective and a holistic approach that provides a promising framework for the realization of proactive urban adaptation measures (Hartmann & Driessen, 2013; Hutter, 2007). These are required to address environmental risks of climate change in cities effectively (Bicknell, Dodman, & Satterthwaite, 2009; Birkmann, Schanze, Müller, & Stock, 2012; Satterthwaite et al., 2009).

Translations into the planning context of the global south

Wilson and Piper (2010) remarked that 'the relationship of planning and climate change is a relatively recent and uncharted territory' (p. 71). In particular, this is still true for the context of the global south. Thus far, the possibility of planning's potentially meaningful role in managing the environmental risks of climate change is mainly informed by research, which, despite an increase in the climate adaptation literature on low- and middle-income countries (IPCC, 2014), remains concentrated on the global north.

Informal urban development reveal that significant parts of urban growth in the global south take place in the shadow of planning and that planning is often limited to reactive measures such as the provision of services after informal occupation of land or relocations. This constrains the effective implementation of non-structural risk management means such as risk conscious land-use management. Accordingly, scepticism that northern planning discourses suffice to 'provide substantial guidance' (Harrison, 2006, p. 328) to scientists, policy-makers and planners in the global south occurs, resulting in warnings of the 'inappropriate "borrowing" of ideas across contexts' (Watson, 2009, p. 151). Informal growth patterns reveal that urban planners in the global south face additional planning challenges that the northern discourse on planning does not address. In addition to planning responses to address general climate change challenges of complexity and uncertainty, issue of efficacy of planning and related planning strategies to respond to informal growth better need to be taken into consideration in the context of the global south.

The urban areas lying between the variable flood lines and the watercourses are, due to regular flooding events, spaces of fluid physical frontiers between land and water. Here, the strong nexus between spatial planning and water management becomes visible, for instance in the provision of space for water retention to manage floods (see Hartmann and Spit, 2014, this issue). The paper provides a contributions to the discussion on land and water governance, which concentrates on exploring possible convergence of spatial planning and water management institutions, two governance fields which despite their nexus have traditionally been rather separated in practice. In the view of informal growth in the global south, solutions are required, which in addition to possible convergence help to prevent informal occupation of flood risks areas and channel future informal growth into safer areas of the city. Through a planning lens, we draw attention to the limits and opportunities of land and water governance in the context of the global south.

Taking urban planning in Johannesburg, South Africa, as an illustrative case and using urban flooding in informal settlements as an example, we explore how responsive planning is to the interrelated planning challenges of informal growth and the environmental risks of climate change. Thereby, we argue that the role that urban planning may in fact play in managing the flood risks of climate change depends on the abilities of planning actors to respond to general and locally specific challenges of climate change and informal growth. Hence, we examine what constrains the realization of adaptive planning options in Johannesburg, a metropolitan city in the global south. We outline contextual lock-in factors that significantly limit the realization of path-breaking planning responses but also demonstrate emerging creative and innovative adaptive planning options addressing informal growth and flood risks of climate change despite contextual lock-in situations.

Path dependency concept: explaining lock-in factors

Literature that deals with the limits of climate change adaptation has evolved mainly around the concept of adaptive capacity (Adger et al., 2009; Carpenter & Brock, 2008) and relates to a broader climate change governance discourse (Birkmann, Garschagen, Kraas, & Quang, 2010; Manuel-Navarette, Pelling, & Redclift, 2011) or to organizational learning and knowledge development (see 'adaptive management'; Allen & Gunderson, 2011; Tomkins & Adger, 2003). It draws an argument around the efficacy of planning – which, however, is significantly reduced in context of informal growth – and speaks to issues of controlling, steering or negotiating urban developments within governance networks and within the framework of available resources and knowledge. Nevertheless, insights gained in the context of the global south relate back to commonly known governance challenges such as capacity constraints or lack of improved means of cooperation, communication and co-production of different actor groups. These governance deficits do not suffice to explain the success or failure of the responsiveness of planning actors. Thus far, key literature has drawn little attention to understanding which urban planning practices of responding to flood risks of climate change are politically feasible and can be justified, and thus legitimized, locally. It requires looking at context factors, such as socio-economic context conditions, as well as societal values and norms (and culture; Garschagen, 2011) in which planning is embedded and in reference to which planning action must be legitimated.

We suggest applying the path dependency concept in order to grasp the interrelation between context factors and the realization of adaptive planning options. Path dependency, basically, is understood as a process continuing to bring about a particular outcome (Sydow, Schreyögg, & Koch, 2005) and that provides a lens through which to examine mechanisms that explain such stability (Kay, 2005). Lock-in situations can be explained as an outcome of path-dependent processes in which a persistent state of specific properties occurs (Vergne & Durand, 2010). Applied to our research, a lock-in would imply that the pool of possible adaptive options which planning actors can in fact realize is substantially limited to a few alternatives. Besides these few alternatives, other options cannot be realized. Although there is a growing interests in understanding adaptation limits – see for instance Holling's (2001) adaptation 'traps' or Adger's (2009) limits of adaptation – the concept of lock-ins has been rather marginalized in research on climate change adaptation. It has found recognition mainly in climate mitigation-related research (Brand & Wissen, 2011), but as well in the planning discourse – although rather marginally (Hartmann, 2012).

Commonly, the increasing returns of self-reinforcing mechanisms *within* an organization or system are used to explain the stability of path dependency and lock-ins, e.g. in the economic literature (Arthur, 1989; Pierson, 2000). However, this system-internal perspective fails to include social contexts and cultural settings, which as well explain stability, for example, of policy (Deeg, 2001; Ebbinghaus, 2005; for 'social expectations', see Sydow et al., 2005). Acknowledging that processes within a system are not autarkic from their context (Luhmann, 1995), the path-dependency concept hence needs adjustments by including consideration of feedback mechanisms between a system and its context. We draw from this adjustment by using path dependency as framework to explore feedback mechanisms of legitimation between the persistence of planning practices and urban planning's context. Two context factors, the developmental context, in which meditation of different planning targets and practices takes place, and the institutional context of collective social norms and values, are looked at specifically for the following reasons.

Metropolitan cities in the global south are prone to a high level of urban divide. As the UN-Habitat (2011b) further puts it:

> cities as diverse as Nairobi, Buenos Aires, Johannesburg, Mexico City, and Rio de Janeiro are similar in that pockets of wealth and poverty co-exist in close proximity. [...] Examples such as these highlight the large disparities between better-off minorities and the many poor, which are also reflected in different degrees of access to [...] facilities, public goods, transportation and open space in most cities in the developing world. (pp. 52–53)

Against this background of urban divide, planning responses to environmental risks of climate change, such as urban flooding, need to be legitimized in the context of complementary or competing development issues. Thereby, this mediation takes place in reference to collective norms and values. Planning is informed significantly by societal norms and values – mirrored in urban vision and planning principles. Being linked mutually to the development context, collective norms and values of a society thus constitute a second frame of legitimacy of persistence in planning practice. This research explores mechanisms of legitimation of the realization (or non-realization) of adaptive planning options, both in the light of justice discourses (what change is legitimized against the background of urban divide and given collective values of equity and justice) and democracy (what change is possible given politicization of urban development and democratic means of resistance to change).

Case study selection and methodology

The research has a qualitative-inductive research design, aiming to provide insights for further research in climate change-related challenges for cities in the global south. The case study approach, with Johannesburg serving as an illustrative case, allows for an in-depth investigation of persistence in planning practices *and* the context that supports this stability.

We selected Johannesburg as case study for several reasons. First, it is at a high risk of flash and river floods caused by heavy precipitation events, especially during South African summer months. The severity of heavy precipitation events has already increased in most parts of South Africa since the 1930s (Mason, Waylen, Mimmack, Rajaratnam, & Harrison, 1999), and is projected to increase further because of climate change (CoJ, 2009). Johannesburg's informal settlements are in particular at a high risk of flooding (CoJ, 2009), with the level of the risk being determined in Johannesburg in accordance to the number of people affected in an area.

Despite the already frequent occurrence of floods in Johannesburg, the risk of flooding does not over-dominate other urban development challenges. In international comparison, floods in Johannesburg can be categorized as relatively small disasters, the devastating consequences of which are primarily due to their cumulative effects (cf. Dickson, Baker, Hoornweg, & Tiwari, 2012). Planning actors in Johannesburg need to negotiate with other urban development challenges that, in public opinion, may be considered more pressing than responses to flood risks of climate change.

This links to the second selection criteria: Johannesburg is one of the most unequal cities worldwide in terms of social polarization and social–spatial disparities; with a Gini coefficient of 0.75 (UN-Habitat, 2011, p. 1/73), it provides an extreme case for the context condition of urban divide. Today's urban divide in Johannesburg is to some extent rooted in apartheid history of various means of racial segregation and exclusion, such as the forced settlement of urban *non-white* population groups in relatively underserviced 'group areas'. While today's dynamics in the city are complex and have in some instances contributed to overcoming apartheid's spatial pattern, forces of globalization are argued to contribute to the development and perpetuation of new lines of urban divide in Johannesburg (Murray, 2011), which are similar to other metropolitan cities, in particular in the global south (Scholz, 2004).

During apartheid, forced evictions and urban planning instruments were used as a strategic tool of racial oppression and segregation. Hence, in the course of apartheid restitution, a strong political link between urban planning and societal values and norms of a socially and ethnically integrated society was built. These norms and values are mirrored in planning visions and strategies of inclusionary urban development and relate to the generation of urban economic growth and the redistribution of targets (cf. Harrison, 2008). Thus, a third reason that Johannesburg was selected was that these historically informed collective norms and values may be an example of a strong contextual frame of reference for legitimation or rejection of adaptive options. In addition, a knowledge gap in South Africa – and in Johannesburg, which constitutes the economic hub and one of the largest and growing cities in Sub-Saharan Africa – on possible adaptive planning options towards issues of climate change further motivates Johannesburg's selection as a case study.

Expert interviews with planning actors serve as the main research method to capture both expert knowledge on site-specific planning contents and the planning actors' subjective perception of what we later categorized as limiting and advancing contextual factors of adaptation of the planning system. Planning actors included state officials within departments of the municipality of Johannesburg (in the following: City of Johannesburg (CoJ)) and CoJ's *quasi-private* entities, and provincial departments that were of relevance to this research, as well as experts in private consultancy companies involved in housing development and risk or environmental assessments. No non-governmental organization (NGO) or community-based organization (CBO) was identified that dealt with flooding specifically in informal settlements. It can be explained by the fact that 'South Africa's planning system is largely state-led, even in a context where the rhetoric of governance and inclusiveness is very strong' (Harrison, Todes, & Watson, 2008, p. ix). Empirical data collection took place during several months between 2012 and 2014. In sum, 23 planning actors were interviewed. Interviews took 1–1.5 hours and were analysed qualitatively. To gain a more complete picture about planning challenges and implemented planning measures, planning documents were analysed qualitatively in addition to the expert interviews, and site visits of state-driven low-income housing projects, slum upgrading projects, and informal settlements in risk areas in Johannesburg were undertaken, accompanied by planning experts.

Johannesburg: housing-based planning approach toward informal growth

Johannesburg has an integrated development planning system, with municipal long-term (Growth Development Strategy Joburg 2040) to medium-term priority plans (the five-year Integrated Development Plan (IDP) and its spatial component, the Spatial Development Framework (SDF)) serving as guidance for urban planning practices and decisions concerning land use and infrastructure planning, services provision, and the related annual budgeting and monitoring process (CoJ, 2010). Thus, urban planning manages urban development in accordance with politically determined development objectives.

Low-income housing used to be an overlapping competence of provinces and municipalities, although the latter often acted as development agents in provincial housing projects, or as a partner in special national projects such as the presidential project of the Alexandra township upgrade. Overall, planners remarked that little integration of provincial housing projects and the IDPs of CoJ occurred, which contributed to CoJ's failure in using low-income housing as means to restructure inefficient urban fabric and contribute to integrated settlement development. In addition to provincial projects, CoJ, however, carried out its own housing projects, such as in the social rental market for upper low-income groups (households with an income of more than 3500 rand per month), or basic service infrastructure projects under the city's informal settlement upgrading policy. Currently, CoJ is undergoing a national accreditation process, implying a shift of central housing competences from Gauteng province to CoJ. Being in the last stage of the accreditation, CoJ is being enabled to manage give-away housing subsidies coming directly from the national government, to act as manager of the housing projects, and, most importantly, it is being tasked with integrating *all* housing projects into its IDPs. It also implies that CoJ can coordinate low-housing housing projects with municipal informal settlement upgrading targets. Together, these make housing an important urban planning tool of integrated development planning in Johannesburg. In the context of land and water governance, integrated development planning therefore constitutes the institutional frame of proactive and long-term flood risks responses in Johannesburg.

Historically informed planning practices

Two legislative frameworks have been advanced that have implications for responding to informal settlements and flood risks of climate change. We briefly introduce them here before outlining their implications regarding flood risks in Johannesburg afterwards. Both legislative frameworks guide planning practice and were developed in reaction to the socio-ethnic and spatial effects of apartheid history.

First, the right to have access to housing is a constitutional right of South African citizens (Constitution of the Republic of South Africa 1996: Art 26(1), 26(2)), motivated by restitution for the legal exclusion of black African, coloured and Indian populations[1] from access to housing opportunities by apartheid legislation. These population groups were restricted to living in ethnically segregated townships that were characterized by limited – since the late 1960s even politically prevented – development of formal housing. In combination with the lack of housing provision, in-migration into these townships contributed to the mushrooming of informal dwellings in backyards and informal settlements, and the overcrowding of formal housing. Consequently, a large housing deficit existed in Johannesburg by the end of apartheid, which affected in particular the black African population. Today, the socio-economic dynamics of globalization contribute to the growth and perpetuation of informal structures in Johannesburg (cf. CoJ, 2012): mainly black Africans, being over-represented amongst the urban poor, remain excluded from access to formal housing by market forces.

The African National Congress (ANC) translated the constitutional 'right to have access to housing' into the Reconstruction and Development Programme (RDP), which 'was eagerly adopted by all political parties and by business' (Harrison et al., 2008, p. 7), and which, in 2005, advanced into the Breaking New Ground Strategy (BNG). Within these legislative frameworks, the state provides single give-away housing units to low-income households with a monthly income of less than 3500 rand, who are in need of formal accommodation, and possess South African nationality. After the first democratic elections such households were captured on the so-called '1996/97 waiting list' for give-away houses. The group of potential beneficiaries has increased since. Today, access to a give-away house is granted to South Africans who fulfil the stated qualification criteria and have not benefitted from a give-away house before, regardless if they are on the 1996/97 waiting list.

Second, the Prevention of Illegal Eviction from and Unlawful Occupation of Land Act of 1998 emerged in response to the Prevention of Illegal Squatting Act (Act 52 of 1951) of apartheid, but also responds to the devastating experiences of many South Africans of being forcibly removed from their homes in the context of spatial–racial segregation during apartheid, with Sophiatown in Johannesburg and District Six in Cape Town being prominent examples. This post-apartheid act implies that evictions of informal dwellers are not possible after 48 hours of informal occupation of land unless the state provides alternative accommodation to evictees.

Recognition: a need to transform planning practices

More than 3 million give-away houses have been provided in South Africa since the first democratic elections in 1994 (NPC, 2012, p. 242). Although the give-away houses practice contributed to improving quality of life for many urban poor, planning actors in Johannesburg call for a transformation of the current housing-based approach towards informal settlements. Their arguments are based in a lack of available capacities and financial resources, which results in failing to keep up housing delivery with the pace of informal growth and reflects the recognition that due to socio-economic developments and the related dynamics of urban divide, informal settlements will remain an integral part of South African cities (cf. Huchzermeyer, 2011). If despite dynamics of persistent informality – the housing backlog in South Africa has grown since the end of apartheid (NPC, 2012) – the formalization practices constitute the main response towards urban informal growth in Johannesburg, planning is likely to remain unable to succeed in addressing interrelated issues of informal growth and flood risks appropriately. Additionally, we will demonstrate that the give-away housing practice contributes to further shaping flood risks in Johannesburg, while also limiting the flexibility of urban planning in responding to the challenges of flood risks and informal growth in a proactive and long-term manner.

Spatial outcome of give-away housing practice further shapes flood risks

A state subsidy is allocated to the qualifying household of a give-away house. This subsidy, however, is handed out to a developer who utilizes it on behalf of the beneficiary by providing the top structure and infrastructure. Due to constraints on subsidy amounts per household, the cheapest building cost per unit is preferred, which translates into the realization of low-density housing designs, thus a large spatial footprint of give-away housing projects. Due to limited availability of (affordable) land in Johannesburg, the most efficient utilization of land is attempted by reducing stand sizes, but also by deviating from standards strongly recommended in South African planning guidelines[2] for the provision of permeable green open spaces and sometimes by constructing give-away houses along (or within) today's 1:100 year flood line (Planner 4, 2013; Planner 19, 2013). In addition, there is strong agreement amongst planning actors that wetland areas and other ecologically sensitive sites are increasingly under pressure to be utilized for give-away housing projects due to scarce affordable land in Johannesburg (Planner 6, 2013; Planner 9, 2013; Planner 11, 2013). Two recent give-away housing projects, Diepsloot East and Malibongwe Ridge, for instance, take place in wetland areas.

Consequently, some areas of give-away housing settlements are at high flood risk. In Orange Farm, water is literally standing inside give-away houses even during minor rainfall events (Planner 7, 2013). Wetlands and the services they provide are increasingly lost: in addition to water purification, they (used to) perform valuable storm water retention functions and, contrary to fixed pipe sizes of hard storm water infrastructure, they would, to some extent, cope better with climate variability in the future. Simultaneously, adequate storm water infrastructure, having to substitute for the loss of wetlands, is either lacking (Figure 2a–c) or of insufficient capacity in give-away housing projects (Planner 1, 2013). Meanwhile, existing systems in other parts of the city are poorly maintained or aged, and thus increasingly fail to perform and compensate (Planner

(a)   (b)

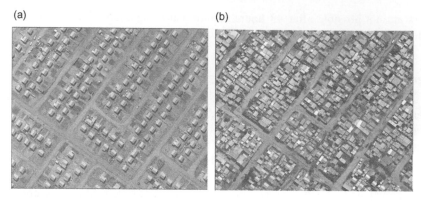

Figure 2.   (a) Give-away housing settlement in Diepsloot West/Johannesburg in 2000 (1:1000), and (b) the same settlement in 2012 (1:1000). Source: Corporate Geo-Informatics, City of Johannesburg.

20, 2013). Consequently, the give-away planning practice shapes current and future urban flood risk in Johannesburg.

Constraints in the realization of proactive and long-term planning options

Forced evictions are hardly desirable or even feasible in view of South Africa's apartheid past. The situation is that there are people living in risk areas, including along flood plains, or in areas which are anticipated to become prone to flooding (e.g. as a consequence of shifting flood lines – due to climate change or siltation of the river), hence they need to be moved to safer sites. In addition, there are cases where informal dwellers need to be relocated in order to upgrade, maintain or replace existing storm water and sewer infrastructure. The Prevention of Illegal Eviction Act protects them from eviction. An option is negotiated relocations and the provision of alternative accommodation. The qualification criteria of give-away houses exclude non-South African citizens and those who have benefited from a subsidy before (e.g. in rural areas), who are estimated to make up a large proportion of informal dwellers. The private formal market does not provide a lowest-income housing product (Planner 4, 2013) and the city's housing alternatives, such as community rental units or formal backyard structures (e.g. in Malibongwe Ridge or Alexandra K 206) require rent payment, which many evictees cannot afford (Planner 8, 2013). Consequently, there is a decoupling between the aim to evict people out of food risk areas and implementation possibilities.

The system of applying a 'housing waiting list' in the allocation of houses complicates negotiations of relocations, since it has to be justified why people (e.g. in flood plains) are housed earlier than community members who, in many cases, have been waiting for a house for a longer time. As a compromise, give-away houses are provided to community members registered on the waiting list, too, in order to be able to evict people out of risk areas or to undertake infrastructure upgrades (Planner 4, 2013; Planner 8, 2013). The Alexandra Renewal Project provides an illustrative example how, consequently, complicated negotiations about the allocation of houses, the urge to provide a larger number of houses than required for those to be relocated, and the search for land suitable for this number of houses, delay the relocation of informal dwellers from flood plains for many years. These factors become a barrier to proactive action, in particular in those cases where the need to evict people under the framework of risk reduction measures is not understood by communities competing over receiving give-away houses who may not support such proactive relocation as a consequence.

Keeping current and possible future risk areas free from occupation is complicated by the 48-hour rule of the Prevention of Illegal Eviction Act, since it would require intensive monitoring of open spaces that, due to capacity constraints, is almost impossible. In Kliptown for instance, '[informal dwellers] have been moved for years. Every year there is a flood, we move them and they [a next generation of informal settlers] come back' (Planner 5, 2013). Besides, several planning actors stated that either they themselves or communities have the impression that informal dwellers purposely settle in flood risk areas to 'jump the housing queue allegedly', anticipating that government is likely to put them in give-away houses if the area floods again. Although only anecdotal evidence is available and further verification of the reported phenomenon is required, it influences planners' perception about the effect (and results in criticism) of current planning practices in addressing issues of flood risks and informal growth. Due to the sum of these constraints in the realization of proactive and long-term measures, planning actors in Johannesburg are missing important opportunities to climate-proof land uses and infrastructure and better manage flood risks today and in the future.

Applied and imagined adaptive planning options

Against the background of the need for change in urban policy and planning, we were interested in which specific changes planning actors imagine, are testing, or have tried to realize to address the interrelated issues of flood risk due to climate change and informal growth. Adjustments in the housing layout to reduce the spatial footprint of give-away housing projects were undertaken, and, in accordance with recommendations of the National Planning Commission, ways to adjust the capital–subsidy funding mechanism of give-away housing (cf. NPC, 2012) are explored in cooperation with the private sector (Planner 5), in order to improve the quality of the public environment, infrastructure provision and location of projects. We highlight three additional changes, which appear to be innovative adaptive options concerning their positive implications for responding to the flood risks of climate change.

Upgrading settlements on informally occupied land

First, the CoJ introduced a three-step programme comprised of two combined policies. As part of the national 'Outcome 8' programme aiming to provide access to basic services to all citizens, informal settlements are serviced with at least water and sanitation regardless if a settlement is prone to flooding or not. Second, the CoJ assesses whether informal settlements are located on risky sites such as flood plains. Dwellers in risk areas are relocated to give-away housing projects. Dwellers in informal settlements, which are suitable for development – in Johannesburg this applies to less than 50% of the informal settlements – can stay and receive title deeds for their stand. Households, which fulfil give-away housing criteria, should receive a give-away house on their stand as well.

We argue that this is a turning point in housing practice in Johannesburg since planning actors are accepting informal dwellings as part of the urban landscape and represents a shift towards planning approaches that better respond to persistent informal growth dynamics. Nevertheless, inappropriate standards of flood risk assessment (e.g. projected climate change variability has not been considered thus far) or attempts to avoid negotiation-burdened relocations may lead to upgrading informal settlements in less-than-optimal locations. Here non-qualifiers in particular, who remain limited to informal housing, are highly exposed to flood risks. In addition, the approach's effects on flood management are likely to be adverse. While provision of infrastructure in informal settlements in risk areas contributes to reducing water contamination during flood events and improved living conditions, it may also attract further in-migration onto these sites. Moreover, its feasibility remains yet to be tested. Given that informal settlements in South Africa are prone to high rates of densification (NPC, 2012), such *in-situ* upgrades become increasingly difficult and may increasingly imply negotiated relocations as well.

Anticipating informal densification

Planners also recognized informal densification in give-away housing settlements since owner of give-away houses rent out backyard shacks on their property (Figure 2a, b). South African census data reveal that the number of backyard shacks has increased to such an extent within the last decade that in 2011 almost 50% of informal dwellers already lived in backyard shacks in Johannesburg. In affected areas, overload of storm water (due to the increase of impermeable surfaces) and sewer infrastructure (due to more households supplied than intended) occurs, which planners hold responsible for sewer overspills and high surface water run-off during and after rainfall events. This leads to damage to formal and informal housing structures and further environmental contamination. Recognizing their limited power in regulating the level of densification in these areas, planners started to discuss the adjustment of infrastructure capacity to informal backyard growth in future give-away housing projects (Planner 4, 2013). This response would allow for proactive planning practices in anticipation of informal growth and in response to related flood risks: first, it addresses an increasingly relevant phenomenon of informal growth by reducing its negative environmental implications; and second, site and services are provided on private properties by market forces, which contribute to channelling informal growth away from informal settlements (e.g. in flood-prone areas) into urban areas that are earmarked to be suitable for development.

Supporting self-help housing on safe sites

Support measures to enhance the building design of backyard shacks were discussed amongst planning actors and have already been adopted in the recent IDP strategy of the city (CoJ, 2012, pp. 53, 83). These translate into the provision of prototype shacks by planning actors, which for instance provide for appropriate ventilation and a stable structure. This self-help housing approach is going to be tested in the township Braamfischerville during the 2013/2014 planning term (Planner 23, 2014). Pilots in five other townships will follow within the next two years. The initiative takes place under the umbrella of the 'Alternative Rental Stock Enablement Programme' and as part of the CoJ's Planning Department's 'Sustainable Human Settlement Urbanisation Plan', a scenario-based plan to accommodate projected future housing demands.

In conjunction with infrastructure adjustment initiatives to respond to anticipated informal densification in give-away housing areas, this programme may provide an attractive alternative for non-qualifiers, at least in those cases where households can afford rent payment for a shack. Nevertheless, the impact of these planning practices is expected to remain limited because initiatives only target non-qualifiers for give-away houses – and non-qualifiers constitute fewer than 10% of informal dwellers in Johannesburg (GDH, 2005).

Contextual lock-in of urban planning practices

Planning actors in Johannesburg perceive and have experienced that shifting away from the provision of give-away houses to qualifiers is not possible due to what we have identified as a contextual lock-in situation. Such transformation of planning practice would require shifting from give-away housing to a housing policy that requires some monetary contribution from those who directly benefit from it. Attempts at incorporating a contribution from qualifiers to their formal accommodation have several possible forms, including rental payment (for rent boycotts, see Wisner, 2009), a financial contribution to construction costs (this was tested in earlier housing project but soon abolished), or the integration of further self-help mechanisms. Second, it would imply redefining who is categorized as a qualifier. This relates to the NPC's (2012) recent view that government should take on 'an enablement role in relation to housing provision [and] shift government spending towards the development of public infrastructure [...] with individuals and communities taking more responsibility for providing their own shelter' (p. 243).

As one planning actor, however, puts it:

> there are many voices going up, even in government, to say that we need to revise our housing policy because it is not sustainable. [...] But we are in a dilemma, because every political party will say, if we are going to say we stop giving [free houses to people], we are going to be voted out. (Planner 2, 2013)

Due to the translation of the constitutional right to access to housing into give-away housing practice, qualifiers anticipate eventually receiving a give-away house, rely on it and act accordingly. For instance, only a few self-help upgrades of informal dwellings can be observed in Johannesburg's informal settlements, despite a relatively high level of security of tenure provided by the Prevention of Illegal Evictions Act. Hence, planning actors find themselves in a dilemma, facing two issues: first, the resistance of those who may not benefit from the policy anymore is expected to confront planners during implementation; and second, planners' challenge of politically justifying why the benefit of a give-away house is not provided as before remains.

Locked in a specific interpretation of collective values

Values of an equal and integrated society are central planning visions in Johannesburg. They are agreed on collectively and 'there is a kind of consensus in this country that we have to continue to focus on social inclusion and social protection' (Planner 14, 2013). If these rather broad norms and values serve as legitimation for planning responses, why does urban planning remain tied to a particular translation of these values into give-away housing practices? Explanation lies in a problematic linkage made between apartheid history of legal exclusion of black Africans from formal urban housing markets and the persistence of the exclusion of many black Africans from formal housing markets through market forces today. Planners report that, therefore, an unquestioned and entrenched sense of expectation to a give-away house amongst low-income black African population is observable, which gains societal legitimization through arguments of affirmative action in the framework of apartheid restitution. Justifying a shift away from give-away housing practice would ultimately imply challenging affirmative action and restitution to the benefit of disadvantaged population groups under apartheid legislation – despite the fact that only 7% of informal dwellers in Johannesburg are actually recorded on the '1996/7 housing list' (GDH, 2005). Contrarily, in political rhetoric, particularly that of Johannesburg's ruling ANC party, a call for even stronger affirmative action is occurring, with service and housing delivery being praised as a means to overcome apartheid's injustices (Zuma, 2014). In addition, other legislation, such as the national affirmative action labour programme for 'black economic empowerment (BEE)' and agricultural land redistribution to the benefit of disadvantaged population groups during apartheid persists, which, again, are politically legitimized by the argumentation of restoring apartheid injustices (Zuma, cited in Makinana, 2014). Shifting planning practice as indicated above would therefore be in conflict with both the content of these legislations, but most importantly with the argumentation of legitimatizing these legislations as well.

Locked in political challenges of urban divide

Path-dependent translation of post-apartheid societal values into persistent legislation shaped a lock-in of urban planning that prevents change in planning practice. However, context conditions of urban divide also contribute to reproducing this lock-in: we suggest that the urban poor use apartheid restitution as a tactic to claim access to specific housing opportunities that they are otherwise unable to realize, and which can be utilized to improve their living conditions. This speaks to the context of urban divide, in which a large population group is excluded from vertical mobilization through formal employment, in particular low-skilled youth, and is likely to be condemned to a life in poverty. Receiving a give-away house means being granted an economic opportunity: planners observe that people move back to their shacks in order to rent out their house (Harrison, 2006, p. 328) or they allow backyard shacks on their premises to gain an income. Changing planning practice as specified above would lead to the abolishment of this economic opportunity provided by receiving a give-away house.

During 'service delivery protests', which often turn violent, the understandable frustration of the urban poor over the slow pace of service delivery materializes, which is linked to the entrenched sense of expectation indicated above.

> There is a serious level of discontent on the ground and we do face protests all the time. It is not the kind of mass protest of the Arab spring, but there are certainly sporadic outbreaks of protests throughout the country, virtually on a daily basis. (Planner 16, 2013)

Protests are triggered even by rather minor intra-community conflicts, but they also mushroom around political elections, since they provide a channel to communicate threats to withdraw votes from leading parties if houses and services are not delivered in the townships where protests are occurring (Runciman, cited in Grant, 2014). These developments indicate that the handing over of give-away houses has become a political instrument to regain and maintain social peace in townships.

Conclusions

Using path dependency as analytical tool, we demonstrated how urban planning practice has locked-in in give-away housing practice in Johannesburg, which considerably limits the realization of novel adaptive planning options towards challenges of flood risks of climate change and informal growth. Realization of planning innovations that emerged in Johannesburg is politically constrained due to expected and experienced resistance to change from population groups that profit from current planning practices by receiving a give-away house. Hence, the application of innovative adaptive planning options is reduced to targeting only a small subgroup of informal dwellers, and consequently the effects of these adaptive options on addressing the interrelated planning issues of flood risk of climate change and informal growth in Johannesburg proactively are conceivably marginal. A nexus of historical paths and current socio-economic context conditions shapes this lock-in.

From a past–present–future perspective, current planning practice is maintained by the identification of restorative justice of post-apartheid society with a particular planning model – the provision of give-away houses – while it arguably continues to increase current flood risks and constrains the realization of proactive and long-term adaptive options towards climate change effects. Although restitution for the ills of apartheid can be achieved through numerous programmes, the South African give-away housing practice became central to popular expectations of a more just social order. Thereby new lines of distributive injustices (Paaval & Adger, 2002) are formed, in the exposure of different subgroups of urban poor, who constitute the most vulnerable population, to flood risks because of this practice today and may further sharpen in future.

Although the history of apartheid is unique to South Africa, elements of this history of exclusion and the aim to undo related pathologies of the past are shared with many post-colonial countries. Comparable feedback mechanism of restorative justice may hence occur in these contexts, which potentially shape similar planning lock-ins. Above all, the lock-in situation in Johannesburg provides an example of a 'Samaritan's dilemma' (Buchanan, 1995) in which the removal of benefits (e.g. of receiving a good free from the state) is politically problematic. Konrad and Thum's (2014) paper on economic policy in climate change adaptation in the European Union reminds us that the consideration of occurrences of lock-ins in a 'Samaritan's dilemma' is important for the development and success of climate adaptation strategies – in the context of both the global south and north.

Taking on a scale perspective, tensions between global changes (such as climate change) and local struggles (such as dealing with issues of urban divide) become apparent. The current planning practice of providing give-away houses supports maintaining the perception that government is able to address the needs of urban poor voters. As such, the Johannesburg case study provides insights into the political limits of climate change adaptation, which are rooted in context conditions of urban divide. A high level of urban divide is observable, in particular, but not exclusively, in many metropolitan cities in the global south. Based on an analytical generalization of the empirical findings with reference to context conditions of a high level of urban divide, the presented findings may hence provide inspiration for further research on contextual limits of climate change adaption in several metropolitan cities in the global south. However, they potentially provide inspiration for research focused on those cities in the global north in which a high level of urban divide also occurs.

Acknowledgement
The authors like to thank Philip Harrison, Ludwig Ellenberg as well as the editors of this issue, Thomas Hartman and Tejo Spit, and an anonymous reviewer for comments and suggestions.

Notes

1. The applied terminology relates back to the categorization of population groups according to ethnic criteria during apartheid, which due to its historical influence on shaping Post-apartheid society is maintained in the societal and academic rhetoric when reflecting on societal dynamics in the country. Hence, research that deals with social problems in South Africa finds itself trapped in applying this terminology.
2. Such as the 'Guidelines for Human Settlement Planning and Design' or principles in the 'BNG Strategy' and the 'Johannesburg Metropolitan Open Space Framework'.

References

Adger, W. N., Dessai, S., Goulden, M., Hulme, M., Lorenzoni, I., Nelson, D.R., ... Wreford, A. (2009). Are there social limits to adaptation to climate change? *Climatic Change, 93*(3–4), 335–354. doi:10.1007/s10584-008-9520-z

Allen, C., & Gunderson, L. (2011). Pathology and failure in the design and implementation of adaptive management. *Journal of Environmental Management, 92*, 1379–1384. doi:10.1016/j.jenvman.2010.10.063

Arthur, W. B. (1989). Competing technologies, increasing returns, and lock-in by historical events. *The Economic Journal, 99*(394), 116–131. doi:10.2307/2234208

Bicknell, J.Dodman, D.Satterthwaite, D. (Eds.). (2009). *Adapting cities to climate change. Understanding and addressing the development challenges*. London: Earthscan.

Birkmann, J., Garschagen, M., Kraas, F., & Quang, N. (2010). Adaptive urban governance: New challenges for the second generation of urban adaptation strategies to climate change. *Sustainability Science, 5*, 185–206. doi:10.1007/s11625-010-0111-3

Birkmann, J., Schanze, J., Müller, P., Stock, M. (Eds.). (2012). *Anpassung an den Klimawandel durch räumliche Planung – Grundlagen, Strategien, Instrumente*. Hannover: ARL.

Brand, U., & Wissen, M. (2011). Die Regulation der ökologischen Krise: Theorie und Empirie der Transformation gesellschaftlicher Naturverhältnisse. *ÖZS, 36*(2), 12–34. doi:10.1007/s11614-011-0031-1

Buchanan, J. M. (1975). The Samaritan's dilemma. In E. S. Phelps (Ed.), *Altruism, morality and economic theory* (pp. 71–86). New York: Russell Sage Foundation.

Bulkely, H. (2013). *Cities and climate change*. London: Routledge.

Carpenter, S. R., & Brock, W. A. (2008). Adaptive capacity and traps. *Ecology and Society, 13*(2), 40.

CoJ. (2009). *Climate change adaptation plan*. Johannesburg: City of Johannesburg.

CoJ. (2010). *Growth management strategy: Growth trends and development indicators*. Second Assessment Report. Johannesburg: City of Johannesburg.

CoJ. (2012). *Integrated development plan 2012–16*. Johannesburg: City of Johannesburg.

Davar, K., Henderson, J., & Burrell, B. (2001). Flood damage reduction. *Water International, 26*(2), 162–176. doi:10.1080/02508060108686901

Davis, M. (2007). *Planet of slums*. London: Verso Books.

Deeg, R. (2001). *Institutional change and the uses and limits of path dependency: The Case of German finance*. MPIfG Discussion Paper. Cologne: Max Planck Institute for the Study of Societies.

Dickson, E., Baker, J., Hoornweg, D., & Tiwari, A. (2012). *Urban risk assessments. Understanding disaster and disaster and climate risk in cities*. Washington, DC: World Bank.

Douglas, I., Alam, K., Maghenda, M. A. *et al* (2009). Unjust waters: Climate change, flooding and the urban poor in Africa. In J. Bicknell, D. Dodman and D. Satterthwaite (Eds.), *Adapting cities to climate change. Understanding and addressing the development challenges* (pp. 201–224). London: Earthscan.

Ebbinghaus, B. (2005): Can Path Dependency explain Institutional Change? Two approaches applied to Welfare State Reform. MPIfG Discussion Paper. Cologne: Max Planck Institute for the Study of Societies. Retrieved from http://edoc.mpg.de/270944

Garschagen, M. (2011). Resilience and organisational institutionalism from a cross-cultural perspective. An exploration based on urban climate change adaptation in Vietnam. *Natural Hazards*. doi:10.1007/s11069-011-9753-4

GDH (2005). *Registration of the informal settlements in Gauteng.* Johannesburg: Gauteng Department of Housing.

Grant, L. (2014). Research shows sharp Increase of Service Delivery Protests. In: *Mail and Guardian*, 14.02.2014. Retrieved from http://mg.co.za/article/2014-02-12-research-shows-sharp-increase-in-service-delivery-protests

Harrison, P. (2006). On the edge of reason: Planning and urban futures in Africa. *Urban Studies, 43* (2), 319–335. doi:10.1080/00420980500418368

Harrison, P. (2008). The origins and outcomes of South Africa's integrated development plans. In M. van Donk, M. Swilling, E. Pieterse and S. Parnell (Eds.), *Consolidating developmental local government. Lessons from the South African experience* (pp. 321–337). Cape Town: UCT Press.

Harrison, P., Todes, A., & Watson, V. (2008). *Planning and transformation. Learning from the post-apartheid experience.* London: Routledge.

Hartmann, T. (2012). Lock-in Situation in Planning: The Role of Law and Property Rights. Presented at the 10th Meeting of the AESOP Thematic Group on Complexity and Planning. Groningen.

Hartmann, T., & Driessen, P. P. (2013). The flood risk management plan: Towards spatial water governance. *Journal of Flood Risk Management.* doi:10.1111/jfr3.12077

Hartmann, T. & Spit, T. (2014). Editorial: Frontiers of land and water governance in urban regions. *Water International, 39*(6), 791–797.

Holling, C. S. (2001). Understanding the complexity of economic, ecological, and social systems. *Ecosystems, 4*, 390–405. doi:10.1007/s10021-001-0101-5

Huchzermeyer, M. (2011). *Cities with slums. From informal settlement eradication to a right to the city in Africa.* Cape Town: UCT Press.

Hutter, G. (2007). Strategic planning for long-term flood risk management: Some suggestions for learning how to make strategy at regional and local level. *International Planning Studies, 12*(3), 273–289. doi:10.1080/13563470701640168

IPCC. (2014). Climate Change 2014. Impacts, Adaptation, and Vulnerability. Working Group II Contribution to the IPCC 5th Assessment Report – Changes to the Underlying Scientific/ Technical Assessment. Chapter 8. Retrieved from http://www.ipcc.ch/report/ar5/

Kay, A. (2005). A critique of the use of path dependency in policy studies. *Public Administration, 83*(3), 553–571.

Konrad, K. A., & Thum, M. (2014). The role of economic policy in climate change adaptation. *CESifo Economic Studies, 60*(1), 32–61. doi:10.1093/cesifo/ift003

Luhmann, N. (1995). *Social systems.* Standford: Standford University Press.

Makinana, A. (2014). Good Story? Zuma chases ANC tale. President calls for intensified affirmative action policy after the elections. In: Mail and Guardian, 21.02.2014.

Manuel-Navarette, D., Pelling, M., & Redclift, M. (2011). Governance as Process. Power Spheres and Climate Change Response. King's College London. Environment, Politics and Development Working Paper Series, Paper 9.

Mason, S. J., Waylen, P. R., Mimmack, G. M., Rajaratnam, B., & Harrison, J. M. (1999). Changes in extreme rainfall events in South Africa. *Climatic Change, 41*, 249–257. doi:10.1023/A:1005450924499

Milly, P., Wetherald, R., Dunne, K. A., & Delworth, T. (2002). Increasing risks of great floods in a changing climate. *Nature, 415*, 514–517.

Murray, M. J. (2011). City of extremes. *The spatial politics of Johannesburg.* Johannesburg: Wits University Press.

NPC. (2012). National Development Plan. Vision for 2013. National Planning Commission.

Paaval, J., & Adger, N. (2002). Justice and Adaptation to Climate Change. Tyndall Centre for Climate Change Research. Working Paper 23.

Pierson, P. (2000). Increasing returns, path dependence, and the study of politics. *The American Political Science Review, 94*(2), 251–267. doi:10.2307/2586011

Satterthwaite, D., Huq, S., Reid, H. *et al.* (2009). Adapting to climate change in urban areas: The possibilities and constrains in low- and middle income nations. In J. Bicknell, D. Dodman and D. Satterthwaite (Eds.), *Adapting cities to climate change* (pp. 3–51). London: Earthscan.

Scholz, F. (2002). Die Theorie der fragmentierenden Entwicklung. *Geographische Rundschau, 54*(10), 6–11.

Sydow, J., Schreyögg, G., & Koch, J. (2005). Organizational Paths: Path Dependency and Beyond. 21st EGOS Colloquium. Berlin, 30.06.2005. Retrieved from http://www.wiwiss.fu-berlin.de/forschung/pfadkolleg/downloads/organizational_paths.pdf

Tomkins, E., & Adger, N. (2003). Building Resilience to Climate Change through Adaptive Management of Natural Resources. Tyndall Centre for Climate Change Research. Working Paper 27.

UN-Habitat. (2011a). Cities and Climate Change. Global Report on Human Settlements 2011. Nairobi: United Nations.

UN-Habitat. (2011b). *State of the World's Cities 2010/2011 – Bridging the Urban Divide*. London: Earthscan.

Vergne, J-P., & Durand, R. (2010). The missing link between the theory and empirics of path dependence: Conceptual clarification, testability issue, and methodological implications. *Journal of Management Studies, 47*(4), 736–759. doi:10.1111/j.1467-6486.2009.00913.x

Watson, V. (2009). Seeing from the South: Refocusing urban planning on the globe's central urban issues. *Urban Studies, 46*(11), 2259–2275. doi:10.1177/0042098009342598

White, I. (2012). *Water and the city. Risk, resilience and planning for a sustainable future*. New York: Routledge.

Wilson, E., & Piper, J. (2010). *Spatial planning and climate change*. New York: Routledge.

Wilson, E., & Piper, J. (2011). *Spatial planning and climate change*. London: Routledge.

Wisner, B. (2009). Environmental health and safety in urban South Africa. In B. R. Johnston (Ed.), *Life and death matters. Human rights, environment and social justsice* (pp. 265–285). London: Left Coast.

Zuma, J. (2014). State of the Nation Address. Cape Town, 13.02.2014.

Land and water governance on the shores of the Laurentian Great Lakes

Richard K. Norton[a] and Guy A. Meadows[b]

[a]Urban and Regional Planning Program, University of Michigan, Ann Arbor, MI, USA; [b]Great Lakes Research Center, Michigan Technological University, Houghton, MI, USA

The Laurentian Great Lakes Basin is large and complex, as is its institutional setting. Given these characteristics, Great Lakes boundaries are both horizontal and fluid, and governance at the Great Lakes water/land interface implicates at least four different frontiers of planning and management. While substantial multinational and sub-national policy regimes have formed over the last century to improve Great Lakes water quantity and water quality management, parallel arrangements have not formed to manage better shoreland boundaries and frontiers.

Introduction

In addition to their ocean coasts, the United States and Canada enjoy tremendous inland seas, the five Laurentian Great Lakes: Superior, Michigan, Huron, Erie, and Ontario (Figures 1 and 2).[1] These lakes and their connecting channels make up the largest freshwater system in the world, extending some 750 miles (1200 km) from east to west and linking the interior of the North American continent to the Atlantic Ocean. Together they represent about 95% of the total freshwater supply in the United States, as well as about 80–90% of North America's and about 18% of the world's freshwater supply. In addition to freshwater consumption, they provide a major natural resource for transportation, industry, power, tourism and recreation.

The combined surface area of all five lakes totals about 95,000 square miles (245,800 km^2), roughly the size of the United Kingdom, and their combined shorelines – including mainland, islands and connecting rivers – totals over 10,300 miles (16,500 km), nearly half the circumference of the earth (GLERL, 2014a; Linton & Hall, 2013; Michigan Department of Environmental Quality (MDEQ), 2014a). In fact, the United States' portion of Great Lakes shorelines, not including islands or connecting waters, totals about 4503 miles (7205 km). Excluding Alaska, that makes the US Great Lakes coastline almost as long as its Pacific, Gulf of Mexico and Atlantic coastlines combined (Gronewold et al., 2013). And while the basin area draining into the lakes is relatively small, totalling about 201,460 square miles (521,830 km^2), it equates nonetheless to roughly the size of the UK and France combined (EPA, 2014a, 2014b; Linton & Hall, 2013).

Given its size and importance, governance of the Laurentian Great Lakes system is highly fragmented jurisdictionally and institutionally as well. Hundreds of units of

Figure 1. Plan-view map of the Laurentian Great Lakes, the Great Lakes states and provinces, and the Great Lakes Basin (watershed).

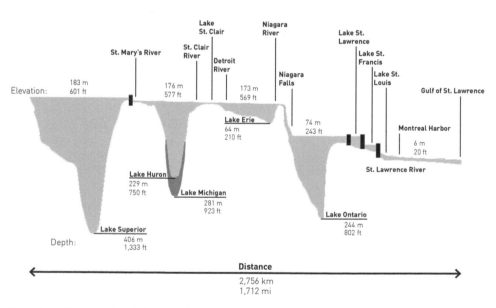

Figure 2. Cross-sectional diagram showing the mean elevations (above sea level) and depths of the five Laurentian Great Lakes, along with connecting waterways and existing dams. Note that Lakes Michigan and Huron are connected hydrologically at the Straights of Mackinac.

government touch Great Lakes shores, including two countries (the United States and Canada), one Canadian province (Ontario),[2] eight US states (Minnesota, Wisconsin, Illinois, Indiana, Michigan, Ohio, Pennsylvania and New York), and numerous localities (e.g., 329 counties, townships, cities and villages in the state of Michigan alone). Beyond these governmental actors, numerous intergovernmental, quasi-governmental and non-governmental organizations have formed over time to influence the management of Great Lakes water and land resources. All these actors engage with one another through a complex and intertwined array of constitutional, statutory and common law legal doctrines, other public institutional structures, and private market forces.

The Laurentian Great Lakes have been surprisingly little studied in terms of shoreland area management despite much recent attention being given to related issues (e.g., Flaherty, Pacheco-Vega, & Isaac-Renton, 2011; Grover & Krantzberg, 2012; Norman, Cohen, & Bakker, 2013). Nonetheless, the Great Lakes Basin provides a useful case for contemplating the challenges of land and water governance at the frontiers of both physical and management boundaries because of attributes unique to that system. Indeed, such contemplation suggests the need to account for both physical boundaries and management systems beyond those proposed by the editors of this special issue. Those same attributes also point to the great contingencies and difficulties that arise when trying to envision and evaluate more integrated land/water governance approaches.

To make out these claims, this paper first presents a brief overview of some of the physical dynamics that make the Laurentian Great Lakes shorelines unique. It then provides a brief overview of the institutions that, taken altogether, comprise the current water and land management system for Laurentian Great Lakes shorelands. Drawing from that background, we contemplate the challenges of integrating land and water governance along Great Lakes shores in terms of understanding both physical boundaries and management frontiers. The scope of this article encompasses the Laurentian Great Lakes system in its entirety, but a focus is made especially on the United States, Lake Michigan and the state of Michigan in particular to provide illustrative examples.

Shoreline dynamics on the Laurentian Great Lakes

Five attributes of the Great Lakes' physical system taken altogether make Great Lakes shores unique compared with ocean–coastal, inland lake and riverine settings, making the challenge of managing land and water on a Great Lakes shore unique as well. The first is that the Great Lakes are geologically young features, with shorelines that are almost uniformly comprised of loose gravels and highly erodible sands, gouged out by glaciers that retreated about 14,000 years ago at the end of the last ice age (Dorr & Eschman, 1970; Rovey & Borucki, 1994). Only very small portions of Great Lakes shoreline are rocky, mostly along the shores of the northernmost Lakes Superior and Huron. As a result, the vast majority of Great Lakes shorelines are highly susceptible to water level fluctuations and other dynamic conditions.

The second attribute is that while the Great Lakes are not tidal, their water levels oscillate naturally on roughly seasonal, decadal and multi-decadal timeframes as a result of changes in precipitation, evaporation, river outflow and groundwater inflow (Gronewold et al., 2013; Meadows, Meadows, Wood, Hubertz, & Perlin, 1997). In addition, 'wind tides' or seiches produce hourly and daily oscillations in water elevations. Figure 3 illustrates the annual average water levels for all of the Great Lakes from 1920 to 2012; while Figure 4 illustrates the annual average lake water levels and the change in

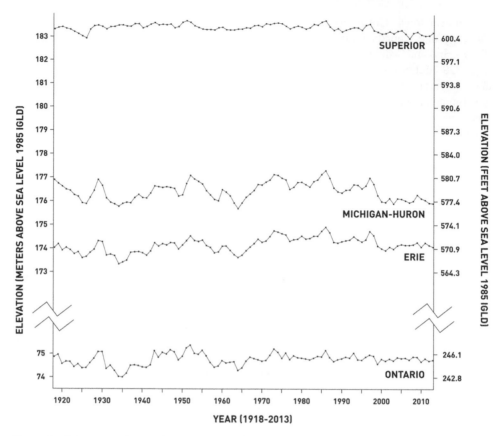

Figure 3.   Mean annual water levels (above sea level) for the Great Lakes from 1918 to 2013. Note that Lakes Michigan and Huron are connected hydrologically at the Straights of Mackinac. The International Great Lakes Datum (IGLD) is a standardized reference elevation, last adjusted in 1985. Source: US GLERL (http://www.glerl.noaa.gov/data/now/wlevels/dbd/).

mean annual lake water levels for Lakes Michigan and Huron (hydrologically connected at the Straits of Mackinac) between 1860 and 2012.

As these figures show, Lake Michigan has experienced horizontal fluctuations in lake water levels of up to 6 feet (about 2 m) over the course of decades. Nested within these annual and decadal fluctuations (but not illustrated on Figures 3 or 4) are seasonal fluctuations that can range on the order of 1–2 feet. Depending on the slope of shoreline, these horizontal changes in water levels can yield vertical shifts in the shoreline of up to tens of feet annually, or hundreds of feet decadally. The remaining three Great Lakes experience similar fluctuations, albeit not as dramatic in general as those changes experienced on Lakes Michigan and Huron (GLERL, 2014b).

It is important to note that of the factors that influence Great Lakes water levels, precipitation and evaporation have by far the greatest effect. Outflows from the system occur, and they have been affected by engineered systems constructed over the past century (including most notably regulation of Lakes Superior and Ontario, the construction of the Chicago Sanitary and Ship Canal, and the dredging of channels connecting the lakes for navigation), but increases in outflow have been offset somewhat by constructed inflows feeding Lake Superior, while inter-lake flows have been moderated somewhat by

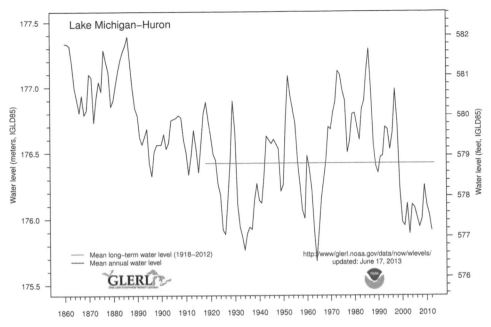

Figure 4.   Mean long-term and mean annual water levels for Lakes Michigan–Huron from 1860 to 2012. Source: US GLERL (http://www.glerl.noaa.gov/data/now/wlevels/levels.html/).

the construction of several dams (Gronewold et al., 2013; Lasserre, 2013). In any event, total outflows amount to only about 1% of the total Great Lakes water budget, such that any changes in water levels caused by engineered systems have yielded only inches of engineered variation compared with multiple feet of natural variation (Buchberger, 1994).

The third attribute that makes the Great Lakes unique is that regional storm patterns affecting the lakes exacerbate shoreline movement resulting from water level fluctuations in several ways. All beaches tend to inflate during the summer season and deflate during the winter (Komar, 1997). These processes are amplified on a Great Lakes shore, however, by increases in storm frequency and severity during periods of rising lake water levels (Meadows et al., 1997) and by basin seiching (sloshing back and forth) during storms. In the long-term, these phenomena result in lake level fluctuations metres above and below the existing mean water level over the course of several hours or a few days (Mortimer, 1987). They also yield changes in the beach morphology of a Great Lakes shoreline over the period of months, seasons and years, often resulting in the appearance of broad beaches under lower water level conditions that appear to be stable and perhaps accreting but that in fact are ephemeral, quickly lost when water levels rise again – possibly even from a single storm event (Bennett, Meadows, Meadows, Caufield, & Sansumeren, 1999).

In addition to shoreline morphology, water level fluctuations and storm patterns, the fourth attribute unique to the Great Lakes is that the entire basin is gradually rising as the earth slowly rebounds isostatically from the weight removed from retreating glaciers. Because the basin is decompressing unevenly and thus causing a shift in the pooling of the lakes generally from north to south (Bennett et al., 1999), this rebound effect has resulted in erosion rates across the basin that vary from what might otherwise be expected, particularly along the southern shores of Lake Michigan.

Finally, Great Lakes water levels always have been and remain a challenge to predict, even on a short-term basis (International Joint Commission (IJC), 2012). But this challenge is yet greater today because of global climate change, despite the substantial empirical and modelling work that has been done, for at least two notable reasons. First, the probabilities used to model the likely future affects of climate on the Great Lakes into the 21st century increasingly lie outside the envelope of historically observed conditions (Intergovernmental Panel on Climate Change (IPCC), 2007), making prediction more speculative. Second, lake water levels are so much affected by both seasonal changes in weather and long-term changes in climate, given their large size and proportionally small drainage basin, that increased uncertainties regarding weather and climate make water level predictions that much more uncertain (Gronewold et al., 2013). Nonetheless, current modelling analyses indicate the potential for stable to modestly decreasing lake levels over time because of climate change, depending on the lake and the modelling scenario used, and the possibility of higher levels at times cannot be dismissed (e.g., Angel, 2013; Angel & Kunkel, 2010; Gronewold et al., 2013). More importantly, no evidence has emerged to suggest that seasonal, annual or decadal fluctuations in lake levels are likely to decrease, even if long-term average water levels do. In fact, variability on many scales appears to be dramatically increasing.

Several important conclusions follow. First, because of the high degree of periodic fluctuations in Great Lakes water levels that occur over time – fluctuations not likely to change despite global climate change – dramatic vertical shifts in the intersection of water and land along Great Lakes shores have occurred and will likely continue into the future (i.e., shorelines will continue to shift dramatically lakeward and landward over time). Most importantly for purposes here, these shifts will continue to yield periodically the apparent accretion of Great Lakes shorelands lakeward, sometimes for extended periods of time. Nonetheless, the long-term trend in most places will continue to be the net loss of shorelands and the progressive landward movement of shorelines as the lakes remorselessly erode them away.

Second, while these horizontal and vertical shifts in Great Lakes shorelines are as extreme, if not more so, than corresponding changes on ocean shores (even given ocean tidal fluctuations and despite global climate change; Gronewold et al., 2013), they play themselves out on an almost slow-motion timeframe – over the course of seasons, years and decades rather than hours and days. Yet problematically, because of water level fluctuations and the other system dynamics just described, not only does the intersection of water and land on a Great Lakes shore shift horizontally over time, but also the profiles of the shoreline beaches themselves change dramatically in response to changing lake levels. Specifically, as noted, low-water-level beaches tend to inflate with sand, making the intersection of the lake with the beach further lakeward than would occur if the beach were comprised of a less moveable substrate. As a result, Great Lakes shorelands that were once submerged may be exposed for years at a time as lake levels decline, while shorelands that have been dry for years may again become inundated – sometimes from a single intense storm event – as lake levels again rise.

The greatest challenge that all these physical dynamics create for land and water governance on a Great Lakes shore is that both shoreland property owners and public officials can easily ignore the long-term consequences of them, at least for a time. Even worse, because water levels periodically decline and shorelines periodically appear to accrete, shoreland users – who invariably want to build as close to the water as possible – are easily lured into a false sense of security, especially during extended periods when lake water levels are low. To be sure, Great Lakes shorelands experience many of the same crises and conflicts typical of ocean, inland lake and riverine systems, including flooding and erosion hazards from short-term storm events and conflicts that arise from competing land uses. But the soporific lull of a seemingly quiescent Great Lake may in reality be a trap that puts private property and public infrastructure at risk in ways not typical of other land/water systems.

Unfortunately, the Great Lakes have experienced a period of relatively low water for going on two decades, and pressures to build closer to them have been increasing. There is every reason to believe lake levels will again rise, perhaps soon, and when they do great risks and harms will follow to all the things that have been built on the land. If ever there was a time to ensure effective and fair land and water governance along the Great Lakes shores, that time is now. Understanding the physical dynamics of Great Lakes shorelines is a necessary step and gets us part of the way, but it is not by itself sufficient. The next task is to understand the institutions that structure land and water governance along Great Lakes shores as they currently exist.

Institutional dynamics within the Great Lakes Basin

Institutionally, the Great Lakes Basin is a multinational, multi-jurisdictional and multi-sectoral system, with a plethora of actors engaging through a broad array of public and private institutional arrangements in a way that makes easy categorization difficult. It is especially difficult to contemplate the array of actors involved separate from the key management issues on which they engage, and vice versa, because most if not all actors engage in most if not all key issues to a greater or lesser extent; actor and action simply do not align in a straightforward way. Nonetheless, to simplify our characterization of the system as much as possible, we first briefly present the key actors involved and then characterize the primary issues engaged, necessitating some overlap between the two discussions. For ease of reference, Table 1 provides a summary of the governmental and quasi-governmental institutional actors noted here.

Key Great Lakes actors

The Great Lakes serve as boundary waters between Canada and the United States. Almost from their very genesis as sovereign nations, these countries began entering into bi-national treaties to resolve specific issues such as disputes over boundaries and the stationing of naval fleets on Great Lakes waters (Crane, 2012). By the dawn of the 20th century, however, the two countries recognized the need to approach Great Lakes water resource management in an ongoing and comprehensive way rather than on an issue-by-issue basis. That recognition led to the signing of the Boundary Waters Treaty (BWT) of 1909, which created the International Joint Commission (IJC) and paved the way for other subsidiary Great Lakes-specific initiatives established in the mid-1950s. These included especially the Convention on Great Lakes Fisheries, which established the bi-national Great Lakes Fisheries Commission (GLFC), and the Great Lakes Basin Compact, which established the Great Lakes Commission (GLC), along with the Great Lakes Water Quality Agreement (GLWQA), signed in 1972 and amended in 1987, which established the Binational Executive Committee (BEC) (Clamen, 2013).

In addition, responding to fears about proposals to export water out of the Great Lakes Basin, the Council of Great Lakes Governors (CGLG) was created in 1983. This council, a partnership of the 10 Great Lakes system states and provinces, adopted the Great Lakes Charter of 1985 in an effort to greatly limit new diversions from the Basin (Crane, 2012;

Table 1.   Governmental and quasi-governmental organizations and their authorities in the Great Lakes Basin related to shoreland area management (presented generally in the order discussed in the text).

Organization	Acronym	Role/programme/authority
International		
International Joint Commission	IJC	Manages Great Lakes transboundary issues pursuant to the Boundary Waters Treaty (BWT) of 1909
		Promotes Great Lakes water quality improvement and protection pursuant to the Great Lakes Water Quality Agreement (GLWQA of 1972, amended 1987; see BEC below)
Great Lakes Fishery Commission	GLFC	Promotes sustainable fisheries management pursuant to the Convention on Great Lakes Fisheries
Great Lakes Commission	GLC	Promotes primarily basin-wide ecosystem management pursuant to the Great Lakes Basin Compact
Binational Executive Committee	BEC	Oversees water quality improvement and protection efforts pursuant to the GLWQA
Council of Great Lakes Governors, and Great Lakes–St. Lawrence River Basin Water Resources Council	CGLC/Compact Council	Negotiates and regulates diversions of water from the Great Lakes Basin pursuant to the Great Lakes–St. Lawrence River Basin Sustainable Water Resources Agreement (2005) and Great Lakes–St. Lawrence River Basin Water Resources Compact (2008)
National		
Environment Canada	EC	Addresses Great Lakes environmental quality under various authorities
		Serves as primary intermediary with the IJC
US National Oceanic and Atmospheric Agency	NOAA	Administers national coastal zone management (CZM) programme
US Federal Emergency Management Agency	FEMA	Administers natural disaster recovery efforts (Stafford Act, as amended) and the National Flood Insurance Program (NFIP)
US Army Corp of Engineers	ACE	Regulates the placement of structures and fill within the federal navigational servitude along, and wetlands adjacent to, Great Lakes shorelines under various authorities
US Environmental Protection Agency/Great Lakes National Program Office	EPA/GLNPO	Addresses Great Lakes environmental quality under various authorities
		Serves as primary intermediary with the IJC
Provincial/state		
Ontario	ON/OMNR	Addresses Great Lakes environmental issues at provincial level through various authorities; delegates and supports shoreland area management authorities through the Ontario Ministry of Natural Resources (OMNR)
Great Lakes States	MN WI IL IN MI OH PA NY	Minnesota (MN), Wisconsin (WI), Illinois (IL), Indiana (IN), Michigan (MI), Ohio (OH) Pennsylvania (PA), New York (NY)
		Address Great Lakes environmental issues at state level through various authorities; promote and support coastal zone management (CZM) programmes pursuant to NOAA; enable local spatial planning, regulatory, and other community-level land management activities

(*continued*)

Table 1. (Continued).

Organization	Acronym	Role/programme/authority
Local		
Multiple local jurisdictions		Numerous counties, townships, towns, cities, villages, and municipalities acting through various legislative bodies (e.g., city councils), administrative bodies (e.g., planning commissions), and administrative officials (e.g., city managers) manage shoreland area development through various delegated spatial planning, infrastructure development, and land-use regulatory authorities and programmes

Lasserre, 2013). That effort subsequently led to the creation of the Great Lakes–St. Lawrence River Basin Water Resources Council (Compact Council, 2014) and the enactment of legislation in all 10 Great Lakes states and provinces that greatly strengthened regional/bi-national control over the use and potential diversion of Great Lakes water (subsequently ratified in the United States by Congress and in Canada by parallel provincial legislation). The IJC, GLFC, GLC, BEC, CGLG and Compact Council thus represent the key multinational governmental actors that exist today to mediate Great Lakes water and land management at the international level.

At the national and sub-national levels, the Canadian Provinces of Ontario and Quebec are the two provinces connected physically to the Great Lakes system, with Quebec bordering the St. Lawrence River and Seaway Basin to the north. The Great Lakes Basin, however, is located entirely within the Province of Ontario. The primary national actor in Canada that addresses Great Lakes water management is Environment Canada, which takes the lead in addressing the country's Great Lakes environmental quality initiatives – primary pollution remediation and ecosystem protection – and serves as the primary intermediary with the IJC (Environment Canada (EC), 2014).

The primary province-level agency addressing Great Lakes shorelands is the Ontario Ministry of Natural Resources (OMNR). This agency does not directly regulate shoreland development. It expanded its efforts to support and improve local planning for shoreline management through its district offices in the early 1990s, however, following an extended period of high lake water levels in the mid-1980s (Lawrence, 1995, 1997, 1998). Nonetheless, it has not taken on additional shoreline management authorities or activities since (OMNR, 2014a).[3] At the regional and local levels, shoreline development is managed primarily by conservation area authorities and local municipalities, including counties, townships and cities, under a variety of provincial policies and statutes, including principally the Ontario Planning Act (OMNR, 2014a; Ontario Environment Commissioner, 2011).

National and sub-national institutions in the United States are both more numerous and more complex. At the national level, four federal agencies play significant roles in shaping Great Lakes shoreland area development. These include: the National Oceanographic and Atmospheric Administration (NOAA), located within the US Department of Commerce, which administers the national Coastal Zone Management Program created by the Coastal Zone Management Act of 1972 (NOAA, 2014); the Federal Emergency Management Agency (FEMA), located within the US Department of Homeland Security, which administers both national natural disaster recovery efforts pursuant to the Stafford Act (FEMA, 2014a) and the National Flood Insurance Program (NFIP) (FEMA, 2014b); the US Army Corp of Engineers (ACE), which regulates the placement of structures and 'dredge and fill' activities within 'navigable waters of the U.S.' (including the siting of structures within the 'navigable servitude' on Great Lakes shorelands and activities affecting wetlands hydrologically connected to Great Lakes waters) (ACE, 2014); and the US Environmental Protection Agency (EPA), which administers national environmental protection laws generally and serves as the primary intermediary at the national level with the IJC through its Great Lakes National Program Office (GLNPO) (EPA, 2014c).

Eight different US states are located within the Great Lakes Basin and touch Great Lakes shores, with the State of Michigan the one state located entirely within the basin and touching four of the five Great Lakes (all but Lake Ontario). Each of the Great Lakes states administers a state-level coastal zone management (CZM) programme pursuant to the federal Coastal Zone Management Act, which did not mandate that states create such programmes but provides a number of incentives to do so (Beatley, Brower, & Schwab, 2002). Each of the state CZM programmes addresses in turn shoreland area management in some form.

The State of Michigan's CZM programme, for example, includes programmes addressing both submerged lands and shorelands that affect near-shore development, pursuant to both federal and state laws and programmes (MDEQ, 2014b). These programmes include state-level permitting requirements that apply to the siting and maintenance of structures within certain near-shore areas, especially state-designated high-risk erosion areas and environmentally sensitive areas.

Beyond the states, numerous localities play substantial roles in planning for and managing the development of Great Lakes shorelands. As noted above, some 329 units of local government touch Great Lakes shores in Michigan alone, and similarly large numbers of local jurisdictions manage shoreland area development in each of the remaining Great Lakes states and the province of Ontario as well.

Finally, beyond this large array of governmental actors, a large and growing array of quasi-governmental and non-governmental actors has appeared over the past several decades to advance particular interests, including the Great Lakes–St. Lawrence River Mayors and the Great Lakes and St. Lawrence Cities Initiative, the Canadian Environmental Law Association, Pollution Probe, Nature Quebec, Great Lakes United, Alliance for the Great Lakes, Council of Great Lakes Industries, the National Wildlife Federation's Great Lakes Regional Office, and the Healing Our Waters-Great Lakes Coalition – just to name more prominent examples (see Crane, 2012, for more detail on all of these groups).

These and other groups formed and have engaged a host of Great Lakes-related initiatives generally as individual entities, typically in response to site- or issue-specific controversies. Even so, many of these groups – including especially the Great Lakes and St. Lawrence Cities Initiative – have increasingly begun to function as a 'transnational municipal network' focused especially on environmental policymaking within the Great Lakes Basin (Kusmierczyk, 2012, pp. 106–107). Most interestingly for present purposes, Kusmierczyk and others assert that these efforts arguably represent a shift away from a 'state-centric, hierarchical, federal-to-federal model' of bi-national relations between the United States and Canada, to one that can be characterized as the 'democratization of bilateral relations [... regarding] transboundary water governance', particularly through the 'internationalization of local government' (Kusmierczyk, 2012, pp. 106–107).

In sum, the actors involved with governance of water and land in the Laurentian Great Lakes Basin are remarkably numerous, diverse and continuously growing, as are the ways in which they are acting and interacting with one another on a variety of related issues over time. It should be noted too that this brief overview of the actors involved, their origins and the kinds of things they do is indeed just that – brief and in no way exhaustive.

Key Great Lakes system management issues

In addition to key actors, water and land governance at their intersection on a Great Lakes shore – that is, Great Lakes shoreland area management – can be organized in terms of the key management issues that have been engaged within the larger Great Lakes system. Moving conceptually and institutionally from water to land to water/land interface, these key issues include ongoing debates over, and approaches to, managing Great Lakes water quantity and quality, coastal community planning and development management in general, and shoreland area development management in particular.

With regard to water quantity, the most prominent concern over the past half-century has been the potential diversion of Great Lakes water to slake the thirst of other regions of the United States and Canada. While this possibility is hard to imagine, the Soviet Union drained much of the Aral Sea in a few short decades to irrigate cotton (Bissell, 2003). So the idea of draining away one or more of the Great Lakes is not out of the question, at least technologically. And indeed there have been grand proposals for substantial diversions from the Great Lakes to serve the Midwest, the arid Southwest, and other regions of the United States since at least the 1950s, while Canada has already engineered substantial cross-basin diversions elsewhere in the country (Lasserre, 2013; Pentland, 2013).

Such diversions from the Great Lakes Basin could have substantial effects on Great Lakes shorelines and shorelands by literally moving them lakeward, if not eliminating them altogether as has happened with much of the Aral Sea. Even so, such diversions are highly unlikely any time soon because of the combination of Canadians' aversion to the idea of diverting Great Lakes water to the United States, the Great Lakes states corresponding aversion to the interstate transfer of water within the United States, the creation of the Compact Council and its authorities to restrict water exports in 2008 (described above), and significant improvements in water conservation measures that have greatly reduced demand for new water supplies across the United States, including its more arid regions (Lasserre, 2013; Pentland, 2013).

With regard to water quality, governance of the Great Lakes system is now widely regarded as a success story for its bi-national, multi-jurisdictional and ecosystem/watershed-based approach (Beatley et al., 2002; Linton & Hall, 2013), for its effectiveness in identifying pollution hotspots and targeting those contaminant sources for remediation (Krantzberg, 2012), and for the extent of collaboration that has been achieved, including – at least arguably – the increased democratization of water quality management (Clamen, 2013; Crane, 2013; Kusmierczyk, 2012). These outcomes have come largely through the bi-national efforts of the IJC, acting under the GLWQA of 1972 as amended in 1987, and through the growing sub-national collaborations described above.

While the Great Lakes are remarkably cleaner now than they were even a few decades ago, a wide array of water quality and biological quality threats still confront the Great Lakes waters today, such as air deposition of contaminated coal ash from power plants (IJC, 2013; MDEQ, 2013). Regarding Great Lakes shorelines and shorelands specifically, water quality issues that remain troublesome include run-off from non-point source pollution (nutrients and sediments), overflow discharges from municipalities with combined storm water/sewer outfalls, persistent contamination from legacy pollution hotspots, and invasive flora overtaking native wetlands species, all of which are degrading consumptive and recreational water uses of near shore Great Lakes waters and Great Lakes fisheries (Healing Our Waters, 2014; IJC, 2013; MDEQ, 2013).

At the other end of the continuum from water governance, governance of land in the United States and Canada generally takes form through land-use (or spatial) planning and development management authorities that originate at the state and provincial level, respectively (see generally Hodge & Gordon, 2014; Levy, 2003). In the United States in particular, there is no coherent or unified national land-use planning policy, such that most planning policies are established at the state and local levels. In the State of Michigan, moreover, as with most states, both local land-use planning and regulatory authorities are broadly enabled and permissive; they are neither mandated nor very constrained or prescriptive (see the Michigan Planning Enabling Act, 2008 Public Act 33, and the Michigan Zoning Enabling Act, 2006 Public Act 110, as amended). As a result, although land-use and community planning in Michigan and elsewhere unfolds within a hodgepodge framework of federal and state infrastructure policies, social laws and policies (e.g., housing assistance programmes), and environmental protection laws, it remains largely a local endeavour (see generally Juergensmeyer and Roberts, 2013). Similarly, most public infrastructure related to land development (e.g., water, wastewater and roads) is provided at the local level, funded substantially by local ad valorem property taxes and fees (see, for example, Citizens Research Council, 1999, for a discussion of local government services and tax authorities in Michigan).

Several attributes of this institutional approach to land governance are important for purposes here. First, because there are so many localities, and because states and provinces have delegated their land-use planning and regulatory authorities so widely and permissively to those localities, especially in the United States, land-use planning endeavours are highly fragmented and uncoordinated. Similarly, because most of these numerous localities are quite small, their financial and institutional capacities to engage in rigorous planning are quite limited (Burby & May, 1997; Norton, 2005a). And for a host of other reasons, not necessarily unique to coastal localities or the Great Lakes, local officials are disinclined to pay attention to their coastal and other unique natural resources through their planning efforts any more than they have to, given all the other community concerns they face. Conversely, a good many communities, at least, are more inclined to worry about promoting economic development and protecting private property rights, whether through their community planning or despite it, than they are inclined to use planning as a means to conserve their natural coastal resources (Beatley, 2009; Beatley et al., 2002; Bille, 2008; Burby & May, 1997; Norton, 2005a, 2005b, 2005c).

Many of these attributes in fact characterize planning as a phenomenon in the United States and Canada generally (the two systems are not identical – but close enough for purposes here). They become unique in coastal settings, and particularly within Great Lakes shorelands settings, because of the unique issues that arise at the water/land interface. In the Great Lakes, these include a range of environmental concerns, economic concerns, the desire to accommodate near-shore residential living, and the need to protect near-shore public and private property from coastal hazards. Key environmental concerns include especially threats to near shore water quality and coastal habitats, discussed above, while key economic concerns include primarily efforts to revitalize working waterfronts and remediate legacy industrial sites, along with efforts to promote tourism through recreational parks, rental beach homes and vacation urban centres (see Ardizone & Wyckoff, 2010, and Pure Michigan, 2014, as examples of these concerns and efforts in Michigan).

Not surprisingly, the fact that these concerns and imperatives are often at odds with one another makes shoreland governance in Great Lakes settings complex and sometimes contentious. What makes them yet more complex – and sometimes even more contentious – are the extra layers of institutional arrangements at play as well. Focusing on the United States, these include most prominently: constitutional laws, common law doctrines and/or statutory laws under which all of the Great Lakes states claim ownership of lands submerged by the Lakes, along with a Public Trust interest in some portion of Great Lakes shorelines – even shores held in private ownership (Norton, Meadows, & Meadows, 2011, 2013; Slade, Kehoe, & Stahl, 1997); federal laws and programmes establishing a 'navigational servitude' along Great Lakes shores that require federal permitting for certain shoreline development and that justify the construction of navigational structures like jetties and harbours (ACE, 2014); federal laws and programmes (FEMA) that map high-risk flooding areas, encourage (and subsidize) flood insurance, mandate local hazard mitigation planning, and provide for disaster recovery following storms (Godschalk, Beatley, Berke, Brower, & Kaiser, 1999; FEMA 2014a, 2014b); and federal (NOAA) and state coastal zone management programmes designed largely to encourage, through policy and financial incentives, more local coastal shoreland area planning and management.

Boundaries and frontiers at the water/land interface on Great Lakes shores

These broad overviews of the physical shoreline dynamics and the institutional dynamics that together comprise the Laurentian Great Lakes Basin provide some insights for rethinking boundaries and frontiers at the water/land interface. The most straightforward insight is that the Great Lakes are best characterized as encompassing two distinct water/land boundaries simultaneously. That is, Great Lakes shores clearly represent a horizontal boundary between water and land as defined by the framing provided for this special issue. They can also be characterized as a fluid boundary, however, one where water and land continuously fluctuate lakeward and landward over years and decades, creating great challenges for managing water and shoreland at any given time – especially when water levels are low for extended periods of time.

In a less straightforward way, the Great Lakes Basin also highlights the need to account for a number or water/land interface management frontiers, which together include but also add to those proposed for this special issue. Figure 5 presents a schematic diagram comparing and contrasting at least four distinct shoreland management frontiers as they have emerged from ongoing Great Lakes management efforts over the past century. This comparison highlights the need to consider not only distinct frontiers conceptually (e.g., spatial planning versus water management planning), but also the key jurisdictional players involved in each (international, national, state/provincial and local), the direction from which management concerns are approached (land or water), the extent to which management solutions tend to be more technical or more institutional, the scale at which the issues are focused (more local to more regional), and the mode of action through which management typically is – or theoretically should be – engaged (more reactive to more proactive).

Consistent with the framing presented for this special issue, there remains a discernible separation in management frontiers regarding Great Lakes shorelands between spatial planning, on the one hand, and water management planning, on the other, at least in the United States. In Michigan, for example, as with most of the United States and Canada, spatial or land-use planning occurs almost exclusively at the local level based on

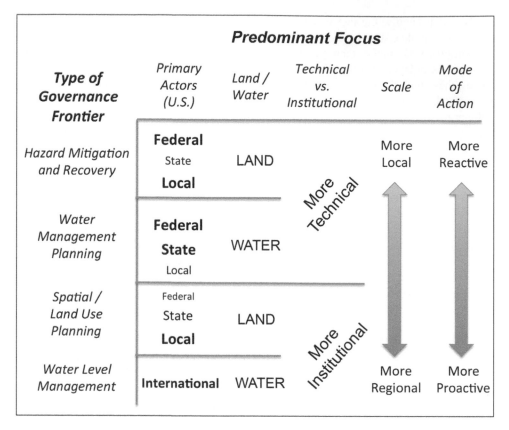

Figure 5. Dimensions of frontiers of governance highlighted by shoreland area management regimes on the Laurentian Great Lakes (focusing on the US system).

permissive and undemanding authorities delegated down from the state. Moreover, contemporary community plans in Michigan's Great Lakes coastal communities tend to focus on generic, community-wide planning issues, conveying surprisingly little attention to their coastal resources in general, let alone demonstrating efforts to coordinate land-use policy making with water management planning in particular (Norton, Meadows, & Meadows, 2009, 2010).[4]

Conversely, conventional water management planning along Great Lakes shores, as elsewhere, has occurred primarily in the form of engineered structures designed to safeguard navigation and protect shoreland properties from storm damage, federal and state-permitting programmes designed to set property development back from high-risk shorelines and/or to strengthen structures built within wave surge and flood hazard areas, and in some cases to beach nourishment in eroding areas. Shoreland navigational and armouring structures are built variously by federal, state and local governments, and to a lesser extent (or at least on a smaller scale) by private property owners (primarily armouring). While these structures are diverse and prolific, one thing that unites them is the remarkable extent to which, historically, their use has been uncoordinated with local spatial or community planning efforts (e.g., Mitsova & Esnard, 2012).

Comparing these two frontiers to one another (Figure 5), water management planning has conventionally been advanced primarily by national and state actors who have approached their task in terms of managing water through technology (i.e., controlling water dynamics using engineered structures), focusing more on the local rather than regional scale (i.e., site-specific) in a generally reactive mode (responding to past damage). Land-use planning has conventionally been advanced primarily by local actors who have approached their task in terms of managing land development institutionally (i.e., through regulations and infrastructure investment policies), focusing more on the community to regional scale in a generally proactive, forward-looking mode. This characterization is broad-brush and not all that remarkable. The important observation, rather, is that this outcome is arguably the result of historical practice and institutional capacities rather than the considered intent to isolate these planning realms (a proposition discussed more below).

Beyond these two frontiers of water/land management, contemplation of the Great Lakes suggests the need to distinguish yet a third planning frontier in coastal settings – hazard mitigation and disaster recovery planning (hereafter, hazard mitigation planning). This mode of planning is actually a mash of two conceptually distinct activities (i.e., planning to prepare for and ideally avoid catastrophic events before they happen and planning to recover from them once they occur), and it overlaps with both water management planning and land-use planning (i.e., when the hazard in question is water-related).

Despite its related focus, and again considering primarily the United States, hazard mitigation planning has evolved over the past several decades as a distinct mode of planning, one engaged at multiple levels typically by actors not well-connected disciplinarily to either water management planning of community planning. At the national level, it takes form primarily through established FEMA flood mapping and flood insurance programmes and more recent federal requirements for county-level hazard mitigation planning. These programmes have prompted in response extensive municipal and county-level hazard mitigation planning efforts. But those efforts, top to bottom, have traditionally been engaged more through civil defence and emergency response institutions and professionals rather than water engineers or community planners (Godschalk et al., 1999).

Numerous calls and efforts have been made in response over the last several decades to better integrate hazard mitigation, disaster recovery, and community planning – often distinguishing them in terms of 'structural' versus 'non-structural' mitigation measures and then noting the need to better coordinate them. Nonetheless, following the long tradition of separation, actual success in achieving such integration has been, to date, wanting (e.g., Mitsova & Esnard, 2012; Puszkin-Chevlin, Hernandez, & Murley, 2006; Peacock & Husein, 2011; US Army Corp of Engineers & Great Lakes Commission, 1999).

Finally, a fourth water/land management frontier that is particularly unique to the Laurentian Great Lakes is the basin-wide water level management regime that has formed over the past century, particularly as it has evolved over the last several years (Flaherty et al., 2011). While focusing primarily on water quantity and then water quality issues – not on shoreland management or for the protection of shoreland properties – long-term management of Great Lakes water levels through control of diversions and other structural systems will have significant impacts on Great Lakes shoreline dynamics into the future. To the extent the lakes can be drained, the technological issues related to diversions and water level management are not insurmountable. Moreover, to the extent that lake water levels can be controlled, most if not all of the structures needed to do so already exist.

Rather than being more technological, therefore, this planning regime falls more within an institutional mode of action, one established at the international levels through the BWT of 1909 and subsequent treaties and organizations, including most notably the Water Resources Compact and Compact Council.

Conclusions and implications for research

The challenges of land and water governance on the shores of the Laurentian Great Lakes implicate the need to recognize that the boundaries between Great Lakes waters and shores are both horizontal and fluid. This characterization is likely consistent with most ocean–coastal settings as well, particularly those susceptible to rising sea levels. Even so, the phenomenon of fluid boundaries on Great Lakes shores is unique and more complicated in the Great Lakes Basin to the extent that Great Lakes shorelines advance and retreat in slow-motion, erratic mimicry of diurnal tidal fluctuations. Especially during extended periods when water levels are low, memories become short and pressures to build on ephemeral sandy beaches grow.

Shoreland area management in the Great Lakes Basin also implicates the need to better integrate not just spatial planning and water management planning, but also to acknowledge the distinct governance frontiers of hazard mitigation planning and Great Lakes water level planning. While planning for water level management is unique to the Laurentian Great Lakes, the need to distinguish hazard mitigation planning from spatial and water system planning is relevant more broadly, at least in the United States. And to the extent that water level planning is already being substantially addressed at the international level in the Great Lakes system, the real challenge is to find ways to better integrate the three remaining frontiers of governance as they exist broadly – water, spatial, and hazard mitigation planning. Can this be done, how might it be done well, and how would we know if it were done well?

Some noticeable success has been achieved in the governance of water quality within the Great Lakes Basin, particularly in terms of the effectiveness of pollution discharge limits and contaminant remediation, as well as in terms of the institutional efficiency and perhaps even the 'democratization' of those efforts, as discussed above. Yet comparable success in terms of land/water governance along Great Lakes shorelands has not been so evident. And it is worth noting that the most compelling challenges that still confront Great Lakes water quality, aside from invasive species, are threats generated by land uses largely under the control of local levels of government – notably non-point source pollution and storm water/sewage overflows from municipal wastewater facilities.

Several factors might explain these divergent outcomes, including especially historical practice and institutional capacities. For political and technological reasons, for example, point-source pollution control in the United States and Canada has been the prerogative largely of national and state/provincial governments, making efforts to more effectively and efficiently coordinate water quality initiatives possible. Conversely, management of Great Lakes shorelands implicates both the regulation of land – a highly contentious proposition – and the need to rely substantially on local governments to do so, units of government that are highly fragmented and uncoordinated and that collectively lack the capacity and inclination to regulate private land uses for environmental or even hazard mitigation purposes. Ongoing efforts to better coordinate and improve these local efforts through incentives, facilitation and encouragement, rather than to reconfigure state-local prerogatives and authorities, speak both to the resilience of these institutional arrangements and the reticence of the states and provinces to reconfigure them.

These observations in fact represent hypotheses to be further explored and tested, particularly in the context of how the management of Great Lakes shorelands might be improved in terms of effectiveness, efficiency, democracy, and justice. Having characterized Great Lakes shoreland governance in terms of the physical boundaries and the management frontiers at play provides a helpful framing for doing so.

Acknowledgements

The authors would like to acknowledge the excellent research assistance provided by Katie Wholey and the helpful comments from Stephen Buckman in preparing this article.

Notes

1. These lakes are referred to as the Laurentian Great Lakes because they comprise a connected system that drains ultimately into the Atlantic Ocean through the St. Lawrence River and Seaway Basin. This system includes connecting rivers and smaller lakes as well, most notably Lakes Nipigon and St. Clair, but the smaller lakes are not commonly referred to as among the five major Great Lakes.
2. The Province of Quebec is also located within the larger Great Lakes system, abutting the St. Lawrence River and Seaway Basin to the north, but it is not located within the watershed that drains into the Great Lakes – that is, the Great Lakes Basin – and does not touch any of the shorelines of the Great Lakes.
3. Aside from brief mention of the ministry's efforts to work 'with local municipalities to help them make wise land-use planning decisions about development along Lake Huron's shorelines' (OMNR, 2014b), no other mention is made of provincial-level efforts to manage Great Lakes shoreline development on the ministry's website, and we have found nothing in the published literature since Lawrence's (1995, 1997, 1998) articles on this topic.
4. These reports are available from the corresponding author upon request.

References

Angel, J. R. (2013). The response of Great Lakes water levels and potential impacts of future climate scenarios. In S. C. Pryor (Ed.), *Climate change in the Midwest: Impacts, risks, vulnerability, and adaptation.* Bloomington, IN: Indiana University Press.

Angel, J. R., & Kunkel, K. E. (2010). The response of Great Lakes water levels to future climate scenarios with an emphasis on Lake Michigan-Huron. *Journal of Great Lakes Research, 36,* 51–58. doi:10.1016/j.jglr.2009.09.006

Ardizone, K. A., & Wyckoff, M. A. (2010). *Filling the gaps: Environmental protection options for local governments* (2nd ed.). Lansing, MI: Michigan Department of Natural Resources and Environment.

Beatley, T. (2009). *Planning for coastal resilience: Best practices for calamitous times.* Washington, DC: Island Press.

Beatley, T., Brower, D. J., & Schwab, A. K. (2002). *An introduction to coastal zone management* (2nd ed.). Washington, DC: Island Press.

Bennett, T., Meadows, L. A., Meadows, G. A., Caufield, B., & Sansumeren, H. (1999). Nearshore profile change and its impact on rates of shoreline recession *Proceedings of Coastal Sediments '99.* Long Island, NY: ASCE (June 1999).

Bille, R. (2008). Integrated coastal zone management: Four entrenched illusions. *S.A.P.I.E.N.S Surveys and Perspectives Integrating Environment and Society, 1*(2), 1–12.

Bissell, T. (2003). *Chasing the sea.* New York: Pantheon Books.

Buchberger, S. G. (1994). *Covariance properties of annual net basin supplies to the Great Lakes.* NOAA Technical Memorandum ERL GLERL-85. GLERL, Ann Arbor, MI.

Burby, R. J., & May, P. J. (1997). *Making governments plan: State experiments in managing land use.* Baltimore, MD: Johns Hopkins University Press.

Citizens Research Council (1999). *A bird's eye view of Michigan local government at the end of the twentieth century.* Livonia, MI: Citizens Research Council of Michigan.

Clamen, M. (2013). The IJC and Transboundary Water Disputes: Past, Present, and Future. In E. S. Norman, A. Cohen & K. Bakker (Eds.), *Water without borders? Canada, the United States, and shared waters.* Toronto, ON: University of Toronto Press.

Compact Council. (2014). *Great Lakes–St. Lawrence River Basin Water Resources Council (Compact Council).* Retrieved from http://www.glslcompactcouncil.org/index.aspx

Crane, T. R. (2012). Great Lakes-Gerat Responsibilities: History of and Lessons in Participatory Governance. In V. I. Grover & G. Krantzberg (Eds.), *Great lakes: Lessons in participatory governance.* Boca Raton, FL: CRC Press.

Dorr, J. A., & Eschman, D. F. (1970). *Geology of the Great Lakes.* Ann Arbor, MI: University of Michigan Press.

Environment Canada. 2014. *Great Lakes.* Retrieved from http://www.ec.gc.ca/grandslacs-greatlakes/default.asp?lang = En&n = 70283230-1

Flaherty, B., Pacheco-Vega, R., & Isaac-Renton, J. (2011). Moving forward in Canada-United States transboundary water management: An analysis of historical and emerging concerns. *Water International, 36*(7), 924–936. doi:10.1080/02508060.2011.628796

Godschalk, D. R., Beatley, T., Berke, P. R., Brower, D. J., & Kaiser, E. J. (1999). *Natural hazard mitigation: Recasting disaster policy and planning.* Washington, DC: Island Press.

Grover, V. I. & Krantzberg, G. (Eds.). (2012). *Great Lakes: Lessons in participatory governance.* Boca Raton, FL: CRC Press.

Gronewold, A. D., Fortin, V., Lofgren, B., Clites, A., Stow, C. A., & Quin, F. (2013). Coasts, water levels, and climate change: A Great Lakes perspective. *Climatic Change, 120,* 697–711. doi:10.1007/s10584-013-0840-2

Healing Our Waters-Great Lakes Coalition. (2014). *Threats.* Retrieved from http://healthylakes.org/threats/

Hodge, G., & Gordon, D. L. A. (2014). *Planning Canadian communities: An introduction to the principles, practice, and participants.* Toronto: Nelson Education.

Intergovernmental Panel on Climate Change. (2007). *Climate change 2007 synthesis report: An assessment of the Intergovernmental Panel on Climate Change.* Geneva: IPCC.

International Joint Commission. (2012). *Lake Superior regulation: Addressing uncertainty in Upper Great Lakes water levels: The International Upper Great Lakes study final report.* Washington, DC: International Joint Commission.

International Joint Commission. (2013). *Assessment of progress made towards restoring and maintaining Great Lakes water quality since 1987: 16th biennial report on Great Lakes water quality.* Ottawa, ON: International Joint Commission.

Juergensmeyer, J. C., & Roberts, T. E. (2013). *Land use planning and development regulation law* (3rd ed.). St. Paul, MN: West.

Komar, P. D. (1997). *Beach processes and sedimentation* (2nd ed.). New York, NY: Prentice-Hall.

Krantzberg, G. (2012). The remedial action plan program, historical and contemporary overview. In V. I. Grover & G. Krantzberg (Eds.), *Great Lakes: Lessons in participatory governance.* Boca Raton, FL: CRC Press.

Kusmierczyk, I. W. (2012). Transnational municipal networks of American and Canadian local governments in the context of bilateral environmental relations: The emergence of a European phenomenon in the Great Lakes Basin. In V. I. Grover & G. Krantzberg (Eds.), *Great Lakes: Lessons in participatory governance.* Boca Raton, FL: CRC Press.

Lasserre, F. (2013). Continental bulk-water transfers: Chimera or real possibility? In A. Norman, A. Cohen & K. Bakker (Eds.), *Water without borders? Canada, the United States, and shared waters.* Toronto, ON: University of Toronto Press.

Lawrence, P. L. (1995). Great Lakes shoreline management in Ontario. *The Great Lakes Geographer, 2*(2), 1–20.

Lawrence, P. L. (1997). Integrated coastal zone management and the Great Lakes. *Land Use Policy, 14*(2), 119–136. doi:10.1016/S0264-8377(96)00039-7

Lawrence, P. L. (1998). Ontario-great lakes shoreline management: An update. *Coastal Management, 26,* 93–104. doi:10.1080/08920759809362346

Levy, J. M. (2003). *Contemporary urban planning* (6th ed.). Upper Saddle River, NJ: Prentice Hall.

Linton, J., & Hall, N. (2013). The Great Lakes: A model of transboundary cooperation. In E. S. Norman, A. Cohen & K. Bakker (Eds.), *Water without borders? Canada, the United States, and shared waters.* Toronto, ON: University of Toronto Press.

Meadows, G. A., Meadows, L. A., Wood, W. L., Hubertz, J. M., & Perlin, M. (1997). The relationship between Great Lakes water levels, wave energies, and shoreline damage. *Bulletin of the American Meteorological Society, 78*(4), 675–682. doi:10.1175/1520-0477(1997) 078<0675:TRBGLW>2.0.CO;2

MI Department of Environmental Quality. (2013). *Michigan state of the Great Lakes 2013.* Lansing, MI: Office of the Great Lakes, MDEQ.

MI Department of Environmental Quality. (2014a). *Shorelines of the Great Lakes.* Retrieved from http://www.michigan.gov/deq/0,4561,7-135-3313_3677-15959—,00.html

MI Department of Environmental Quality. (2014b). *Coastal Management.* Retrieved from http://www.michigan.gov/deq/0,1607,7-135-3313_3677_3696—,00.html

Michigan, P. (2014). *What's New in Pure Michigan.* Retrieved from http://www.michigan.org/

Mitsova, D., & Esnard, A. -M. (2012). Holding back the sea: An overview of shore zone planning and management. *Journal of Planning Literature, 27*(4), 446–459. doi:10.1177/088541 2212456880

Mortimer, C. H. (1987). Fifty years of physical investigations and related limnological studies on Lake Erie, 1928–1977. *Journal of Great Lakes Research, 13*(4), 407–435. doi:10.1016/S0380-1330(87)71664-5

Norman, E. S. Cohen, A. & Bakker, K. (Eds.). (2013). *Water without Borders? Canada, the United States, and Shared Waters.* Toronto, ON: University of Toronto Press.

Norton, R. K. (2005a). More and better local planning: State-mandated local planning in coastal North Carolina. *Journal of the American Planning Association, 71*(1), 55–71. doi:10.1080/01944360508976405

Norton, R. K. (2005b). Local commitment to state-mandated planning in coastal North Carolina. *Journal of Planning Education and Research, 25*(2), 149–171. doi:10.1177/0739456 X05278984

Norton, R. K. (2005c). Striking the balance between environment and economy in Coastal North Carolina. *Journal of Environmental Planning and Management, 48*(2), 177–207. doi:10.1080/0964056042000338145

Norton, R. K., Meadows, G. A., & Meadows, L. A. (2013). The deceptively complicated 'elevation ordinary high water mark' and the problem with using it on a Laurentian Great Lakes shore. *Journal of Great Lakes Research, 39,* 527–535. doi:10.1016/j.jglr.2013.09.008

Norton, R. K., Meadows, L. A., & Meadows, G. A. (2009). *Lake level dynamics in Michigan's Great Lakes: Implications for shoreline management policy and law (Phase I interim research report).* Prepared for the Michigan Department of Environmental Quality, Lansing, MI.

Norton, R. K., Meadows, L. A., & Meadows, G. A. (2010). *Lake level dynamics in Michigan's Great Lakes: Implications for shoreline management policy and law (Phase II final research report).* Prepared for the Michigan Department of Environmental Quality, Lansing, MI.

Norton, R. K., Meadows, L. A., & Meadows, G. A. (2011). Drawing lines in law books and on sandy beaches: Marking ordinary high water on Michigan's Great Lakes shorelines under the public trust doctrine. *Coastal Management, 39*(2), 133–157. doi:10.1080/08920753.2010. 540709

ON Ministry of Natural Resources. (2014a). *Great Lakes.* Retrieved from http://www.mnr.gov.on.ca/en/Business/GreatLakes/index.html

ON Ministry of Natural Resources. (2014b). *Lake Huron.* Retrieved from http://www.mnr.gov.on.ca/en/Business/GreatLakes/2ColumnSubPage/STEL02_173901.html

Ontario Environment Commissioner. (2011). Land use planning in Ontario: Recommendations of the Environmental Commissioner of Ontario from 2000–2010. Retrieved from: http://www.eco. on.ca/blog/2011/01/25/land-use-planning-in-ontario-ten-years-of-eco-recommendations/

Peacock, W. G., & Husein, R. (2011). *The adoption and implementation of hazard mitigation policies and strategies by coastal jurisdictions in Texas: The planning survey results.* College Station, TX: Hazard Reduction and Recovery Center, Texas A&M University.

Pentland, R. (2013). Key Challenges in Canada-US Water Governance. In E. S. Norman, A. Cohen & K. Bakker (Eds.), *Water without borders? Canada, the United States, and shared waters.* Toronto, ON: University of Toronto Press.

Puszkin-Chevlin, A., Hernandez, D., & Murley, J. (Winter 2006/2007). Land use planning and its potential to reduce hazard vulnerability: Current practices and future possibilities. *Marine Technology Society Journal, 40*(4), 7–15. doi:10.4031/002533206787353141

Rovey, C. W. I., & Borucki, M. K. (1994). Bluff evolution and long-term recession rates, south-western Lake Michigan. *Environmental Geology, 23*(4), 256–263.

Slade, D. C., Kehoe, R. K., & Stahl, J. K. (1997). *Putting the Public Trust Doctrine to work* (2nd ed.). Washington, DC: Coastal States Organization.

US Army Corp of Engineers, & Great Lakes Commission (1999). *Living with the lakes: Understanding and adapting to Great Lakes water level changes*. Detroit, MI: US Army Corp of Engineers, Detroit District.

US Army Corp of Engineers. (2014). *Great Lakes and Ohio River Division*. Retrieved from http://www.lrd.usace.army.mil/

US Environmental Protection Agency. (2014a). *The Great Lakes: An Environmental Atlas and Resource Book*. Retrieved from http://epa.gov/greatlakes/atlas/index.html

US Environmental Protection Agency. (2014b). *Great Lakes: Basic Information*. Retrieved from http://epa.gov/glnpo/basicinfo.html

US Environmental Protection Agency. (2014c). *About the Great Lakes National Program Office (GLNPO)*. Retrieved from http://www2.epa.gov/aboutepa/about-great-lakes-national-program-office-glnpo

US Federal Emergency Management Agency. (2014a). *Robert T. Stafford Disaster Relief and Emergency Assistance Act*. 2104a. Retrieved from http://www.fema.gov/media-library/assets/documents/15271

US Federal Emergency Management Agency. (2014b). *The National Flood Insurance Program*. Retrieved from http://www.fema.gov/national-flood-insurance-program

US Great Lakes Environmental Research Laboratory. (2014a). *About Our Great Lakes: Great Lakes Basin Facts*. Retrieved from http://www.glerl.noaa.gov/pr/ourlakes/facts.html

US Great Lakes Environmental Research Laboratory. (2014b). *Great Lakes Water Level Observations*. Retrieved from http://www.glerl.noaa.gov/data/now/wlevels/levels.html

US National Oceanic and Atmospheric Administration. (2014). *Coastal Programs: Partnering with States to Manage our Coastline*. Retrieved from http://coastalmanagement.noaa.gov/programs/czm.html

Index

Note: Page numbers in *italics* represent tables
Page numbers in **bold** represent figures
Page numbers followed by 'n' refer to notes